WATER POLLUTION AS A WORLD PROBLEM

Water Pollution as a World Problem

the Legal, Scientific and Political Aspects

REPORT OF A CONFERENCE HELD AT
THE UNIVERSITY COLLEGE OF WALES,
ABERYSTWYTH

11/12 July 1970

Chairman
THE RT. HON. LORD HODSON, P.C., M.C.

Published for

The David Davies Memorial Institute
of International Studies

By

EUROPA PUBLICATIONS
LONDON

EUROPA PUBLICATIONS LIMITED
18 Bedford Square, London, W.C.1

ISBN   0 900 36241 3
Library of Congress Catalog Card Number 71-154484

JAPAN
Maruzen Co. Ltd., Tokyo

AUSTRALIA AND NEW ZEALAND
James Bennett (Collaroy) Pty. Ltd., Collaroy,
N.S.W., Australia

INDIA
UBS Publishers' Distributors Pvt. Ltd., Delhi 6

Printed and bound in England by
STAPLES PRINTERS LIMITED
at the Stanhope Press, Rochester, Kent

Contents

List of Participants   vii

Saturday, 11 July 1970

Chairman's Opening Address
The Rt. Hon. Lord Hodson, P.C., M.C.   1

Nuclear and Thermal Waste Pollution   11
Chairman: Mr J. E. S. Fawcett, M.A., D.S.C.
Professor E. H. S. Burhop, B.A., M.Sc., Ph.D., F.R.S. (Scientist) –
Nuclear Waste Pollution   13
Mr A. Preston, B.Sc. (Administrator) – *Nuclear and Thermal Waste
Pollution*   19
Dr D. W. Bowett, M.A., LL.D., Ph.D. (Lawyer) – *Nuclear and
Thermal Waste Pollution: Some Legal Considerations*   22
Discussion   27

Oil Pollution   51
Chairman: Professor P. F. Wareing, D.Sc., F.R.S.
Professor R. B. Clark, Ph.D., D.Sc. (Scientist) – *The Biological
Consequences of Oil Pollution of the Sea*   53
Mr. Maurice Holdsworth A.M.I.E.E., M.I.M.E. (Administrator)–
Oil Pollution: Industry   73
Professor R. Y. Jennings, Q.C. (Lawyer) – *Oil Pollution: the Law*   77
Discussion   83

Sunday, 12 July 1970

Chemical and Pesticide Pollution   101
Chairman: Dr D. W. Bowett, M.A., LL.D., Ph.D.
Mr M. Owens, B.Sc., M.I.Biol. (Scientist) – *Chemical and Pesti-
cide Pollution*   103
Mr N. J. Nicolson, B.Sc., A.R.I.C. (Administrator) – *Adminis-
tration in the Control of Chemical and Pesticide Pollution*   112
Professor J. F. Garner, LL.D. (Lawyer) – *Chemical and Pesticide
Pollution: the Law*   115
Discussion   121

Industrial Waste and Sewage Pollution   141

Chairmen: Sir Frederick Warner, B.Sc., Hon.D.Tech., D.Sc.,
     C.Eng., F.I.Mech.E., M.Cons.E.
     Lord Simon, C.M.G.

Mr A. J. O'Sullivan, B.Sc., M.I.Biol. (Scientist) – *Pollution by
Industrial Waste and Sewage: Scientific Aspects and Some
Problems of Control in the Marine Environment*   143

Mr R. F. Pearson, B.Sc., A.C.G.I., D.I.C., F.I.C.E., M.I.W.P.C.
F.I.P.H.E. (Administrator) – *The Greater London Council's
Work in the Control of Water Pollution*   152

Mr I. Drummond (Lawyer) – *Pollution by Industrial Waste and
Sewage: Legal and Administrative Aspects*   160

Discussion   168

Appendix I

President Nixon's Message on Ocean Dumping to the Congress of
the United States, October 1970   205

Appendix II

Speech of Hon. Mitchell Sharp, Canadian Secretary of State for
External Affairs, introducing the Arctic Waters Pollution Bill,
in the Canadian House of Commons, April 1970   207

Text of Arctic Waters Pollution Bill   213

Appendix III

Covering Note of International Wildfowl Research Bureau Convention on Wetlands of International Importance especially as
Wildfowl Habitat   229

Text of Convention on Wetlands of International Importance   235

Aberystwyth, Wales, 11-12 July 1970

Participants

DR W. K. ALLAN, Dept. of Mechanical Engineering, University College, London.

DR W. J. ANTHONY-JONES, Director of General Studies, University College of Wales.

MR B. M. ARCHIBALD, Chemical Industries Association, London.

MR PATRICK ARMSTRONG, Parliamentary Group for World Government, London.

PROFESSOR D. R. ARTHUR, Dept. of Zoology, University of London, King's College.

MR M. J. BEASLEY, Howard Humphreys & Sons, Engineers, Surrey.

MR W. O. BINNS, Forest Research Station, Farnham, Surrey.

DR W. R. P. BOURNE, The Seabird Group, Watford.

DR D. W. BOWETT, Queens' College, Cambridge.

DR G. BROWN, Postgraduate School of Physics, University of Bradford.

MR J. R. BUCKENHAM, The British Waterworks Association, London.

PROFESSOR E. H. S. BURHOP, Dept. of Physics, University of London.

DR JEAN CARROZ, Fisheries Dept., FAO, Italy.

MR CARTER, Office of the Chief Public Health Inspector, Warrington.

PROFESSOR R. B. CLARK, Dept. of Zoology, University of Newcastle upon Tyne.

THE HON. JONATHAN DAVIES, Llandinam, Montgomeryshire.

DR C. G. DOBBS, School of Plant Biology, University College of North Wales.

MR IAN DRUMMOND, Clerk of the Trent River Authority.

PROFESSOR E. EISNER, Applied Physics Dept., University of Strathclyde.

MR J. G. EVANS, Richards & Dumbleton, Engineers, Birmingham.

MR M. R. EVANS, The West of Scotland Agricultural College, Glasgow.

MR J. E. S. FAWCETT, Director of Studies, Royal Institute of International Affairs, London.

DR J. D. FISH, Dept. of Zoology, Aberystwyth University.

MR P. S. G. FLINT, Imperial Chemical Industries, London.

MR M. J. GAINES, Technical Writers' Section, United Kingdom Atomic Energy Authority, London.

PROFESSOR J. F. GARNER, Dept. of Law, Nottingham University.

DR J. R. GOLDSMITH, Dept. of Public Health, State of California, USA.

MR G. GOODWIN, Ministry of Technology, London.

PROFESSOR IVOR GOWAN, Dept. of Political Science, University College of Wales.

THE EARL OF GOWRIE, London.

MR J. B. GREENWOOD, Chamber of Shipping of the United Kingdom, London.

MR J. C. HANBURY, Allen & Hanburys Ltd, Herts.

Dr G. P. Hekstra, Secretary, Koninklijke Nederlandse Akademie Van Wetenschappen, Biologische Raad, Netherlands.

Dr P. R. Helliwell, Dept. of Civil Engineering, Southampton University.

The Rt. Hon. Lord Hodson, Rotherfield Greys, Oxon.

Dr A. J. Holding, Dept. of General Microbiology, University of Edinburgh.

Mr Maurice Holdsworth, Shell Petroleum, London.

Dr J. W. Hopton, Dept. of Microbiology, University of Birmingham.

Dr Gwyneth Howells, Natural Environment Research Council, London.

Mr E. H. Hubbard, Stitchting Concawe, Netherlands.

Mr G. W. Hull, Commonwealth Human Ecology Council, Surrey.

Professor R. Y. Jennings, Whewell Professor of International Law, Jesus College, Cambridge.

Mr A. K. Jones, Dept. of Botany, University College of Wales.

Mr F. G. B. Jones, Dept. of Biology, Brunel University, London.

Mr A. Neville Jones, Water Quality and Fisheries Dept., S.W. Wales River Authority.

Mr B. G. Little, Land Resources Division, Directorate of Overseas Surveys, Surrey.

Mr F. Macdonald, Monsanto Chemicals Ltd., Newport.

Mr J. McLoughlin, Faculty of Law, University of Manchester.

Mr R. Maitland Earl, National Pure Water Association.

Dr W. G. Marley, Radiological Protection Division, UK Atomic Energy Authority, Berks.

Dr G. V. T. Matthews, Wildfowl Trust, Slimbridge.

Dr J. G. Merrills, Faculty of Law, University of Sheffield.

Dr David M. Miller, Counsellor, Canadian High Commission, London.

Professor A. Newell, Jordans, Bucks.

Professor Lily Newton, Aberystwyth.

Dr G. Nickless, Dept. of Inorganic Chemistry, University of Bristol.

Mr N. J. Nicolson, River Purification, Thames Conservancy, Berkshire.

Mr A. J. O'Sullivan, Lancashire and Western Joint Sea Fisheries Committee.

Mr M. Owens, Water Pollution Research Laboratory, Stevenage.

Mr R. F. Pearson, Dept. of Public Health Engineering of the Greater London Council.

Mr Niel Piercy, Centre for Environmental Studies, London.

Mr A. Preston, Fisheries Radiobiological Laboratory, Lowestoft.

Mr A. Pride, Shell International Petroleum Co., London.

Mr I. D. Richardson, White Fish Authority, London.

Dr K. E. W. Ridler, National Research Development Corporation, London.

Mr J. Roburn, Office of the Government Chemist, London.

Mr T. Schofield, Leicester City Engineer.

Viscount Simon, Port of London Authority.

Mr Walter Simon, Europa Publications, London.

Dr T. N. Stevenson, Dept. of the Mechanics of Fluids, University of Manchester.

Mr J. R. Stribling, Chief Inspector, Eastern Area, Thames Conservancy.

Mr J. J. Swift, London.

Mr Eric Taylor, Wirral, Cheshire.

Dr J. A. G. Taylor, Unilever Research Laboratory, Cheshire.

Dr J. B. R. Taylor, School of Biological Sciences, Bath University of Technology.

Mr D. H. Thomas, Rechem Ltd., Southampton.

Mr A. R. Tindall, Dept. of Physiology, University of Birmingham.

Mr A. H. Walters, The Soil Association, Suffolk.

Mr P. Walters Davies, The Nature Conservancy, Wales.

Professor P. F. Wareing, School of Botany, University College of Wales.

Sir Frederick Warner, Cremer & Warner, Engineers.

Dr Gillian White, Faculty of Law, University of Manchester.

Mr J. F. Whitfield, Chairman of the Pollution Prevention Committee, Severn River Authority.

Professor H. T. Williams, Dept. of Agricultural Economics, University College of Wales.

Dr E. Windle Taylor, The British Waterworks Association, London.

Mr A. S. Wisdom, Avon and Dorset River Authority.

Mr D. G. Wood, Manchester.

Mr R. E. Woodward, Mersey and Weaver River Authority.

Dr R. J. Wootton, Dept. of Zoology, University College of Wales.

Professor P. A. H. Wyatt, Dept. of Chemistry, St. Andrews University.

Mr S. L. D. Young, IMCO, London.

Chairman's Opening Address

CHAIRMAN: THE RT. HON. LORD HODSON, M.C.

Once again I would like to begin by welcoming to the Conference all those who, by their presence here, show that the subjects to be dealt with are of sufficient interest and importance to induce them to give up a summer weekend to attend these proceedings. We are indeed delighted to have such a distinguished array of scientific, legal and administrative talent with us here but, in opening the Conference, I think I should stress that, from the scientific angle, I speak as a layman. Moreover, I assume that quite a number of people here are as innocent as I am about technical matters and some of you are very proficient, particularly in your own fields. This Conference will, I think, be most valuable in bringing together both sides in order that not only the specialist but also the general public may, somehow, be better informed on the problems about which we are going to talk.

I am asked to say that it would be very convenient if all the speakers, when they come to their turn to speak, would give their names, and, secondly, perhaps this affects me more than many others, if elaborate technical language is used it might be a good idea to translate it. I have been reading some of the papers and quite a number of the nouns employed were not in the dictionary and no doubt some, both adjectives and nouns, have been invented by some of you who are here today. So it may be necessary to come down from your great heights and explain what the range of capital letters and the more esoteric terms employed stand for.

When we convened our first Conference on Law and Science in 1964, we were indeed breaking new ground. Since then concern over the damage which technological and scientific advance has increasingly inflicted upon our environment has steadily grown. Much of our awareness today is due to the work of Rachel Carson, whose warning of the dangers of the continuing widespread use of modern chemical pesticides in her book *Silent Spring* has recently been endorsed in substance by no less an authority than Dr Kenneth Mellanby. In a recent article in *The Times* he describes it as one of the "most important ecological phenomena of our time".

He goes on to say, "The world has many problems. The population explosion, increased industrialisation of and changes in agriculture could all cause pollution which, at least, would destroy the quality of civilised life and, at worst, could destroy life itself." Curiously enough, this phrase about industrialisation and pollution caused by industrialisation was in my mind yesterday when I bought a copy of *Punch*. I don't know whether any of you have seen the cartoon there, a picture of an industrialist standing beside his factory, pouring effluent into the river, with a student standing by his side. The industrialist says to him: "That may be effluent to you, my lad, but to me it's a new Bentley and a heated swimming-pool."

1

That draws attention to the question of the necessary balance which arises in connection with these questions as those of you who have read the Helsinki Rules dealing with the pollution of rivers will realise. It isn't enough to say, "All pollution is bad" because all industrial work, or nearly all, seems to involve some pollution; the question is, "How much pollution can we stand?" That is, I suppose, the practical problem which confronts the business world; it doesn't very much interest the idealist who would like to get rid of pollution altogether.

Our purpose here is to deal in some depth with one particular aspect of pollution – water pollution. Perhaps I may be forgiven if I refer here to Article 9 of the Helsinki Rules which defines the term "water pollution" as referring to "any detrimental change resulting from human conduct in the natural composition, content or quality of the waters of an international area". That, of course, is dealing with rivers but the marine definition on the same lines, defined by a body called GESAMP (that is a joint group of experts from IMCO/FAO/UNESCO/WMO) as "the introduction by men, directly or indirectly, of substances or energy into the marine environment including estuaries, resulting in such deleterious effects as: harm to living resources, hazard to human health, hindrance to marine activity, including fishing, impairment of quality for use of sea water and reduction of amenities".

But, in opening our deliberations, I make no apology for setting the subject in its wider context by drawing attention to some of the more generally publicised matters connected with pollution. For, in fact, pollution of the environment is *one* subject, indivisible save for convenience when attempting to deal with its various aspects.

There are numerous examples of pollution which come readily to mind. Air and water are inextricably mingled so must be considered together. Sulphur dioxide from the unwashed plume of smoke from the Battersea power station is precipitated by rain. In the Pennines the Forestry Commission has abandoned a project to re-afforest the slopes near the industrial areas since the sulphur dioxide kills the trees; the Scandinavians allege that contaminated wind-borne rain from our Midlands is killing the fish in their lakes and the Norwegians complain that pollutants carried by wind from the Ruhr turns their snow black and, in due course, the snow melts and runs into rivers and lakes. Recently Dr Frank Taylor, of the Institution of Heating and Ventilating Engineers, has said that the amount of sulphur dioxide in the atmosphere may rather be increasing than decreasing.

In Canada, the great Queen Elizabeth Highway, surrounding the "golden mile" of industry round Toronto, is ice-free throughout the year. But its condition is due to the use of salt and the Canadians are now finding that the run-off is rendering the adjacent fields barren and poisoning their rivers in the vicinity – there is, of course, an historical precedent for this, in the case of William the Conqueror sowing the Vale of York with salt with disastrous results which lasted for many years.

Not so long ago, the waters of the Norfolk Broads were so clear that you could see the bottom and study its vegetation – now nearly everywhere the water is turgid for most of the time due to the thick growth of algae resulting from the wash-off of fertilisers from the land and the discharge of phosphates in household detergents. Hickling Broad is particularly badly affected in this way.

The condition of the Baltic and North Seas is giving rise to considerable anxiety and their capacity for absorbing further, and indeed increasing amounts of wastes is considered to be, at the best, doubtful. The Soviet Union is seriously concerned about the state of the Caspian Sea which, in a recent official report, is described as "being choked up and poisoned". It "receives discharges of oil, petroleum products, industrial and city sewage, ballast and waste from ships". The level of the Sea has fallen 8 ft. in the last forty years and the number of fish has sharply decreased. The same report lists equal damage to Lake Baikel. It goes on to catalogue an overall shortage of fresh, usable water in the Ukraine, steppe-lands blowing away in dust storms, the erosion of invaluable shore-lines and the increase in chemically laden smoke over cities. Pollution is already serious in fifteen rivers, including the Don and the Volga, and it has been accompanied by what is described as "the mass slaughter of fish" – something presumably akin to the recent disaster on the Rhine and the very much smaller incident here reported on 20 June when hundreds of fish were found dead in Chichester Yacht Harbour two days after the aerial spraying of crops with phorate granules. The great Kuznets coal mines are said to be "devouring fertile land, poisoning the air and polluting the water". The stage would seem to be set for extensive anti-pollution legislation by the new Supreme Soviet.

It was recently reported that the Venetian canals, carrying sewage flushed twice a day by the tide, were full of dead fish and decaying seaweed. Nearer home the problem of the disposal of sewage in the Black Country has been described by the River Trent Authority as appalling. The Trent itself is so polluted that this year the annual raft race has had to be cancelled, while the Tame which flows into it is said to be the most heavily contaminated river in Britain.

According to the annual report of the Association of Public Analysts, published last year:

"Indiscriminate discharges have converted many of the rivers of northern England and the lower reaches of the Thames into biological deserts. The seas into which they flow are in danger of falling into the same state unless rapid and firm action is taken."

The fact that Los Angeles is "choking to death" on its own fumes is too well known to need much emphasis although some of the more esoteric details may not be familiar. The air is so polluted that radios broadcast smog bulletins at regular intervals giving the precise pollution count in the atmosphere. When the level of danger is reached people at risk are warned to stay indoors and schools are forbidden to carry out physical exercises in the open until it falls. When there is no wind the exhaust fumes from approximately five million cars can be seen to rise upwards like mist "into the thick blanket of garbage [i.e. carbon dioxide] that hangs suspended over the city". The car manufacturers have so far been unwilling to fit pollution controls on cars as it would probably cut sales and profits – but the ultimate choice lies, of course, with the consumer and there are signs of an approaching change. The California State Legislature had indeed laid down that all cars should cease to run on leaded petrol by January 1971, but

3

in the event it bowed to pressure and shelved the Bill. In Tokyo the majority of the inhabitants habitually wear masks to eliminate fumes. In addition, they hang cages of white mice in the streets – when they die the situation is held to be "temporarily serious". The Greek Chamber of Technology recently informed the authorities that as a result of an investigation carried out at their request, they had come to the conclusion that unless the ever-increasing air pollution was dealt with as a matter of urgency Athens would have to be abandoned in ten years' time.

To turn for a moment to pollution of the sea which is, as far as the high seas are concerned, largely a question of oil pollution. This, of course, is not the case with the off-shore and estuarine waters where pesticide residues, industrial waste and sewage play a greater role. Where oil is concerned you already have the beginnings at least of some international control. Moreover, I think it fair to say that the major oil companies, which, of course, have vast international empires, are very conscious of their responsibilities and have themselves taken steps to deal with the threat in the arrangements under the Load-on-Top agreement and the TUPALOV insurance scheme. IMCO has the situation under constant review and progress has been made as a result of the conclusion of the International Convention for the Prevention of the Pollution of the Sea by Oil in 1954, amended in 1962, and the Convention on the High Seas in 1958. It is convening a further international conference in 1973 "for the purpose of preparing a suitable international agreement for placing restraints on the contamination of the sea, land and air by ships, vessels and other equipment operating in the marine environment".

But the fact remains, however, that there are constant reports of oil spillages at sea; for instance early this year thousands of sea birds and some seals perished off the coasts of Northumberland, and a slick threatened the Suffolk coast, again resulting in very heavy bird deaths. Almost all our beaches have oil residues washed up on to them, including the Scillies and the Channel Islands. The same phenomenon is common throughout Europe and the Americas. For instance, in May a seven-mile oil slick was reported off the South Devon coast. During June, a British tanker sank off the Seychelles in the Indian Ocean, spewing tons of fuel oil into the sea 6 miles from the main port of Victoria, and Thor Heyerdahl reported to Dr Hambro, Norwegian Ambassador to the United Nations, that throughout twenty-seven days sailing on *Ra* he had observed 1,400 miles of oil spreading from horizon to horizon "as far as the eye could see in all directions in the mid-stream current of the Atlantic". He later delivered samples of the oil lumps, resembling solidified asphalt, to the UN oceanographic ship *Calamar*, which confirmed his findings. When the Greek tanker *Arrow* struck a rock off Cape Breton shores she spilt some one million gallons of heavy bunker C oil which, in the cold water, some 39°F., quickly turned into something that is almost impossible to handle, is not biogradable and will not burn. In addition to this, several hundred miles of Denmark's North Sea coast was threatened by an oil slick only two weeks ago. Recently up to a million sea birds have died along an arc on the western side of the Alaska peninsular, the probable cause being oil.

So far the greatest cause of nuclear pollution of the sea comes from the fall-out from weapon tests, but the use of nuclear power sources for oil drilling and

4

pumping on the sea-bed is actively under consideration. No central register or means of assessing what amounts of nuclear waste, when and in what condition, have gone into the sea exists, but in 1967 a group of European states (Belgium, France, Federal Germany, the Netherlands and the United Kingdom), together disposed of 11,000 metric tons of radioactive waste in an undertaking organised by the European Nuclear Energy Association.

Another possible cause of pollution is the case of an accident of the type which has already occurred at least twice, when the Windscale reactor went critical in 1957, emitting considerable amounts of radioactivity, more than was released at Hiroshima, and when the Enrico Fermi breeder reactor in the USA did the same in 1966 endangering Detroit. On the first occasion a fortuitous inversion of wind dispersed the radioactivity harmlessly, in the second they were able to contain it and shut the plant down. But the man-made levels of radiation are bound to rise slowly over the next decades as the use of reactors spreads and the chances of engineering failure also rise *pari-passu*.

Very little study has yet been made of the long-term cumulative effects of concentrations of low radioactivity. But there is some evidence available on this subject. In a study of the Columbia River in the western United States – reported in *The Atom and the Energy Revolution* – radioactivity was shown to accumulate progressively in a remarkable manner. In reports prepared by J. J. Davies, R. W. Parker, R. F. Palmer, W. C. Hanson and J. F. Cline for the Biology Operation, Hanford Laboratories, Richland, Washington, for the First and Second United Nations Conferences on the Peaceful Uses of Atomic Energy in June 1955/58, entitled *Radioactive Materials in Aquatic and Terrestrial Organisms Exposed to Reactor Effluent Water*, it was stated, *inter alia*, that there was "a tremendous accumulation of different radio isotopes in organisms. . . . Many of the isotopes found in the organisms were not measured in the river water because they occurred in amounts below detectable limits in the 2 litres of water that were tested." For one such isotope, Zinc 65, it was said: "Although its abundance in the organisms is less than phosphorous sodium or iron, the Zinc 65 was readily transferable through the food web and occurred in relatively large concentrations in almost all organisms sampled." They went on, "The importance of the food web upon the accumulation of radio isotopes by organisms" was demonstrated by the results of the experiments. It was found "that sucker fish were the most radioactive at the second tropic level, they fed directly upon sessile algae. . . . Fission products released into the atmosphere in gaseous or particulate form, or released as process waste water, enter terrestrial biological chains principally by deposition on vegetation, absorption into plants from substrata, and ingestion or inhalation by animals. Of particular interest to contamination potential is the ability of waterfowl to effectively concentrate several of the fission products when chronically exposed to a source of the material."

"When aqueous solutions of industrial wastes containing relatively minor amounts of fission products are impounded, plant succession may make the resulting swamps attractive to waterfowl. Of the fission mixture, radioisotopes of strontium, cesium, ruthenium and several of the rare earths have sufficiently long half-lives to make them available to plants and animals for extended periods. Strontium is usually the most important member of this group because it is

5

significantly absorbed by plants and animals and because of its low permissible body burden. Although the concentration of strontium in vascular plants is less than 20 per cent of the concentration in the soil in which they are grown, the feeding habits of waterfowl utilising such a swamp will effect transfer of applied amounts of radiostrontium to the bird."[1]

They conclude: "Of the many radioisotopes that have been measured in Columbia River water, none were present in quantities near the recommended maximum limits for drinking water by the International Commission on Radiological Protection. Studies in which rats were maintained on reactor effluent that contained concentrations of radioisotopes several thousand times the concentrations in the river water indicated that the danger of assimilating harmful levels of radioisotopes by drinking Columbia River water was negligible. Significant quantities of various isotopes were concentrated, however, in the bodies of most organisms which had access to the contaminants via the natural food web. It is apparent that food organisms of man could accumulate hazardous levels of certain radioisotopes from water which contained concentrations of the contaminants that were well within the permissible limits for drinking water. Representative organisms are routinely sampled from natural environments which are exposed to dilution of effluent from the Hanford reactors to ascertain that the amounts of radioactive contamination are well below hazardous levels.

"The data presented emphasises the value of the need for radio-ecological research to further elucidate and evaluate environmental relationships that pertain to radioactive waste disposal."[2]

Radioactive waste is either sunk at sea or buried in steel tanks, many of which contain highly active waste and have to be constantly cooled. Of 183 such tanks located in Washington, South Carolina and Idaho, nine have already failed and their contents have had to be transferred after less than twenty years. Moreover, tanks of highly active waste have been buried in a region known to be subject to earthquakes. No substance known so far to man can contain such waste for the thousands of years that some of it will remain active and, in the light of the Columbia River investigations, how radioactive is the water discharged from reactor power stations used for fish breeding?

One of the most controversial issues at the moment has been that raised by the discovery of oil in Alaska. On Wednesday, 22 April 1970, the Canadian House of Commons voted to adopt Prime Minister Trudeau's proposal, embodied in the Arctic Water Pollution Bill, to extend Canadian territorial waters from 3 miles to 12 and to exercise pollution controls in Arctic waters 100 miles out to sea.

The crucial point was reached as a result of the attempts of the huge United States tanker *Manhattan* during the past year to prove that the Northwest Passage could be tamed by modern marine technology. Escorted by Canadian icebreakers it has done so with qualified success. Most of the Northwest Passage is within 100 miles of the Canadian mainland and, at one point of the route followed by the *Manhattan* last summer, there is a neck barely 6 miles across. The Humble Oil Company of America planned to run a year-round service of

[1] UN Doc.: A/Conf.8/P281, July 1955.
[2] UN Doc.: A/Conf.15/P393, June 1958. Experiments conducted in May 1970 confirmed the findings reported earlier. *AEC Research and Development Report.*

6

icebreaking super tankers through the Northwest Passage to the big markets on the American east coast. Canada will also insist that all vessels operating within 100 miles of the Canadian shoreline must be subject to Canadian regulations drawn up with the protection of the environment in view. Canadian national interests and environmental conservation can be said at this point in time to run hand-in-hand.

The Arctic is the source of thermal and weather influences which affect the whole of the globe. We are, at present, quite unable to assess how serious the effects of sinking a tanker carrying 60 million gallons of crude oil would be. There are three main elements in the situation: the scale of the operation imposes the risk of accidents which would make the *Torrey Canyon* and the Santa Barbara blow-out look very small; the proposed service would be operated through extremely dangerous waters where accident is more likely than in any other seas, and the coldness of the Arctic waters means that oil does not break up or evaporate and the damage done would be semi-permanent. The natural processes of biogradability that, in other parts of the world, would in time clean up even a huge oil spill do not work in such cold weather. Canada is also taking other steps to preserve the Arctic. A caterpillar tractor carelessly run across the fragile tundra will so dislocate the thermal regime of the permafrost that a huge area can be turned into a lake within a few months. To meet with Alaska trucking interests a road was built to the Prudhoe Bay oilfields; it was in fact gouged out of the tundra. A year later it was simply a vast bog. If the proposed pipeline TAPS is carried over or through the tundra it would, of course, be liable to leaks. In 1968 there were 200 pipeline leaks in the United States in pipes much smaller than that proposed for the Arctic, 100 of which were spills of between 1,000 and 12,000 barrels. The results of such a spill under Arctic conditions can be imagined. So far, no plans for the construction of the road or pipeline have been approved.

Finally, the United States has protested at the steps taken by the Canadian Government, saying that it did not recognise any exercise of coastal jurisdiction over "our vessels on the high seas" and thus would not recognise the right of any state unilaterally to establish a territorial sea of more than three miles offshore, and suggested that the dispute should be sent to the International Court of Justice. Mr Trudeau replied in a speech that there was no law on pollution controls and declared, "We will not go to Court until such time as law catches up with technology." In fact, the lack under existing international law of regulatory powers applicable prior to an accident was one of the reasons advanced by the Canadian Government in justification of its decision to establish direct unilateral control over shipping entering Arctic waters. The main legal framework for the sea is becoming increasingly inadequate; national law is fairly effective, but international rules, save for oil where they are still often evaded, are vague and without sanction.

Intensive dredging, equivalent to strip-mining, in particular areas could do long-lasting damage to the flora and fauna of the marine environment. Fears of such possible damage were expressed, as reported by *The New York Times*, on 6 April, over a project to mine 52 million acres off the Bahamas for aragonite. Considerable damage has already been done off the Indonesian coast by mining

7

and, as has been said in a recent study, "knowledge of the effect of dredging and associated processes on the benthonic biological regimes (flora and fauna of the bottom of the sea) and their susceptibility to environmental change is almost completely unknown". Since marine drilling techniques are in advance of means for restraining or removing oil pollution if a blow-out, *vide* Santa Barbara, or other incident occurs, the conclusion is reached, in the words of the United States delegate to the United Nations, that "the chance for accidents of massive proportion in this environment is a very real one".

As I indicated a little earlier, we have to look at conservation of the environment from both points of view since trouble may arise from the most *prima facie* beneficial operations. For instance, the Aswan Dam will not only vastly increase the agricultural yield in Egypt, but it has also already had the effect of destroying the shrimp fisheries at the mouth of the Nile and of making the Mediterranean considerably more salt in that area. It has also meant the spread of bilharzia into new areas. It further entails the import of vast quantities of fertilisers and pesticides if the new types of high yielding grains which are now being planted are to flourish. This again entails pollution of the water by run-off from the land.

The increasing need for the establishment of international standards for permissible agricultural and industrial residues is illustrated, *inter alia*, by the fact that Scotch rivers are now in danger of pollution from Australian pesticides. So much pesticide is in use in the pastures that Australian wool is heavily impregnated – the imported wool washed in Scotland for spinning contaminates the rivers.

"Efforts to improve the environment in Britain are at best mediocre and at worst criminally negligent", according to the Chief Public Health Inspector of Warrington, Mr E. Ward, in an article in *Municipal Engineering*. Conservationists are aghast at the "self-congratulatory pronouncements of politicians and officials, that much progress has been made to improve the environment, particularly in the field of air pollution". The industrial areas of the North and the Midlands are falling behind; smoke pollution is better but pollution by sulphur dioxide and motor vehicles continues. He adds that tests carried out at Warrington to measure lead deposits suggest that California, in spite of its smog and motor vehicles, has less pollution than Warrington. The Standing Technical Committee on Synthetic Detergents of the Ministry of Housing and Local Government has reported that increasing use of hard detergents has halted improvement and has again increased foaming on the rivers after considerable improvement.

In the United States "dispose first and investigate later" seems to be the rule, "an invitation to disaster that requires no documentation for the proof of sinister changes in the estuarine life of many coastal areas in the United States is dismally at hand for anyone to examine". A sevenfold increase in the industrial wastes disposed of in the seas over the next decade is forecast in a recent publication, *The Physical Resources of the Ocean*. Over a million square miles of shell-fish producing waters bordering the United States are now unusable and there has been a very great increase of pollution over the past few years. Where international rivers are concerned the Helsinki Rules on the Uses of the Waters of International Rivers at least lay down that "waters of a given drainage basin are

8

regarded as an integrated whole and not as a series of separate entities wherein each State may proceed as it wishes".

Irreversible ecological damage has probably been done in America where pesticides were initially used with such abandon. In the tropics ecological damage is still being done but many pests are becoming increasingly resistant to DDT so that its useful life is limited. As Dr Mellanby says, "Scientists throughout the world are now aware of the dangers of pesticides and are developing safer chemicals, and are devoting more attention to developing non-chemical or biological methods of control". But, even so, care is still necessary. Paraquat, so recently pronounced as "safe", has just been proved to have poisoned hares who came into contact with it when sprayed from the air.

"Today", C. L. Sulbergh writes, "one can almost reckon that the more advanced a nation is, the more polluted its soil and the more poisoned its enveloping atmosphere." There are, however, signs that a more cautious attitude is growing. The United States Secretary of the Interior, Mr Walter Hickel, has recently imposed a virtual ban on DDT, Aldrin, 2,H,5-T, Dieldrin, Endrin, DDD, all mercury compounds, and nine lesser known agents in more than 500 million acres of federal land. He has placed thirty-two other chemicals and classes of chemicals on a "restricted list"; they are allowed to be used under limited circumstances but only with the approval of the Cabinet Committee on Pesticides. In one country the use of the organochlorines is being phased out but not so far as I am aware, their export. On 8 June the Oil Corporation announced that it would halt production of DDT, of which it produces about 20 per cent of that manufactured in the United States, on 30 June this year. The announcement came three days after three conservation organisations filed suit in the Federal District Court in Washington that Olin be enjoined from discharging wastes containing DDT into waters leading into the Wheeler National Wildlife Refuge in Alabama.

West German industrialists are reported to be setting aside 6 per cent of their new plant investments for the control of pollution, while, on 12 May, the French Government announced wide-ranging plans to conserve the environment.

On 11 June, President Nixon asked Congress to cancel twenty federal oil leases in the Santa Barbara Channel area, the firms concerned to be compensated. The area thus effectively closed off to further drilling is about 198,000 acres 20 miles off the Santa Barbara coast. He has further set up a centre in Washington to co-ordinate the efforts of six federal agencies for handling oil spills. On the other hand, the long-term effects of the defoliation policy pursued as part of the Viet-Nam offensive over a period of several years are so far quite incalculable. For the moment, the use of defoliants has been halted.

These are, of course, only a few of the very many steps that are either being taken or contemplated for dealing with environmental problems.

To turn our attention once again nearer home we are all aware of how very greatly the water of the Thames has been improved in recent years. The Thames Conservancy is to spend £6–8 million over the next ten years on a pumping system, as I understand it from the reports which have appeared, which will be capable of adding 100 million gallons of water a day to the river. To do so they will draw on the natural reserves in chalk and limestone regions as underground reservoirs, they will be utilised during dry periods and allowed to fill up during the winter. Very careful systems of calculation and analysis will be employed in deciding what can safely be done without permanently depleting water resources. I think the Thames Conservancy stands high in reputation for its attitude and the work it is doing in connection with water pollution.

In concluding this necessarily very superficial and incomplete survey, which is merely intended as a perhaps useful background, I would like, if I may, to make one or two more general remarks. I do so in all humility because I don't pretend to be an expert in this or, indeed, in any particular subject. In the first place, it is quite obvious that one of the more pressing needs is for internationally accepted standards in all the fields on which I have touched. For not only is pollution of the environment an indivisible whole, in terms of its component parts, it is also inescapably a global problem. Air and water know no national frontiers. Both investigation of the causes and effects of pollution and control measures must be "global in scope and scientifically inclusive in range of enquiry". And, moreover, such investigation and assessment embracing all the disciplines involved, must be carried out before new techniques are employed or new projects adopted, and not after the damage has been done. We cannot rely upon science and technology to remedy the results of our abuse of science. If we have learnt nothing else it must surely have been borne in upon us that applied scientific techniques, while solving the problem with which they have been confronted, more often than not in doing so present us with further unforeseen problems which become progressively harder to solve if we have ignored their existence in the first instance.

In the second place, conservation of the environment and particularly con-servation of amenities will be a very expensive business. But, in the long term, it may well prove even more expensive not to take steps to safeguard man's environment. To advocate our eventual flight to other planets as a means of survival would seem, at this juncture, a council of despair. It should, however, be stressed that the choice to be made and the measures to be taken are *essentially* political and only at the advisory level, scientific or technological.

Finally, in the wider context it may perhaps be worth considering whether it is really possible to envisage a world in which all the inhabitants, wherever situated, can attain a Western European type of industrialised civilisation. It would seem on the face of it that two choices may, at some point, have to be made. Whether, under all circumstances, resources should be exploited even when such exploitation may entail total destruction of a given environmental area and, in the second place, whether there are not certain more difficult terrains in which the choice may be either the continuation of what we describe as a more "primitive" way of life or the abandonment of that area as being uninhabitable.

But such speculation is to go too far beyond the subject of our present deliberations.

I now call upon Mr Fawcett, as Chairman of our first panel on nuclear and thermal waste.

The David Davies Memorial Institute of International Studies | The Department of International Politics of the University College of Wales, Aberystwyth

CONFERENCE ON

LAW, SCIENCE AND POLITICS: WATER POLLUTION AND ITS EFFECTS CONSIDERED AS A WORLD PROBLEM

Saturday 11 July

MORNING SESSION

Nuclear and Thermal Waste Pollution

Chairman MR J. E. S. FAWCETT
Director of Studies, Chatham House

Scientist PROFESSOR E. H. S. BURHOP
London University

Administrator MR A. PRESTON
Fisheries Laboratory, Lowestoft

Lawyer DR D. W. BOWETT
President, Queens' College, Cambridge

MR J. E. S. FAWCETT (Chairman of 1st Session on Nuclear and Thermal Waste Pollution)

I do not wish to delay the beginning of the first discussion, but I want to say this: that however slight the acquaintance may be that anybody makes with the problem of pollution it serves to show that, even in the authoritative writing on the subject in books and journals, an immense amount of contradiction exists. I think the purpose of this Conference could very usefully be to clear up a great deal of the confusion between fact, rumour and speculation in these various fields, and I hope we shall succeed in doing this.

As regards the first subject what I would hope to do is to ask the speakers to talk for some 15 to 20 minutes each, followed by a few questions limited to clearing up points of fact in their statements, and then we would go on to the full discussion. Professor Burhop.

centrations of fission products and dispose of the remainder into the normal effluent, hoping that it contained no radioactivity. But one cannot make a complete separation in this way – it is not as simple as that. One has to dissolve these radiated rods and go through various chemical processes and when one has made a solution the volume of material to be dealt with is very great indeed. So one starts to try and concentrate it by evaporation and ideally one would like, as I said, eventually to get to a stage where all the radioactivity is in one small part and all the rest completely harmless. This is not possible, but it has been possible, with the practices which have been used, to reduce the amount that has to be dumped into the sea by a factor of the order of 10^5 – one hundred thousandth of the radioactive material.

To my mind the real problem is not that which is put into the sea but that which is left behind. At the present time, a great deal of research is going into the question of the disposal of this residue of extremely radioactive material. With such a high concentration of fission products, where energy is being given out the whole time, it has to be cooled to prevent it from boiling and, in fact, it is cooled by passing water pipes around it and it is kept at a temperature of about 60°C. Then it is placed in stainless-steel tanks of extremely high-quality construction and these themselves are encased in walls of concrete about 8 ft thick. Provision is made always to have spare tanks around for an emergency. Should a leak develop in one of the storage tanks it can be pumped away into one of the reserve stainless steel tanks, while even if the storage tanks should be destroyed catastrophically, it is hoped that a stainless-steel lined concrete shell would take the material which comes out of the inner stainless-steel tank. No doubt very many precautions have been taken and, I understand, there have been no failures in twenty years but, of course, as I said before, these tanks have to last not twenty years but thousands of years. One has to deal with this problem and our descendants will have to deal with it for thirty to forty generations ahead. The engineering problems involved here are very great and serious ones. This is one of the questions that perhaps Dr Marley, or somebody else from the Atomic Energy Authority, might enlarge upon and say how they envisage this problem of these very highly concentrated radioactive effluents – not those put in the sea, but those left behind – being dealt with.

I know that a great deal of research is going on in trying to convert these concentrated effluents into solid glass-like materials which can be more easily stored. I do not know what the situation of that research is; whether it looks promising and whether this will prove to be the solution.

We were much impressed with the care that had been taken by the scientists of the Atomic Energy Authority, and also by the scientists in Dr Preston's laboratory, over this question of what happens to the tiny amount of radioactive material which is deposited in the sea. Lord Hodson has referred to the concentration of radioactive isotopes in different stages of the various biological cycles. The researches which were made into the effluent in Cumberland showed that, using the critical path technique, or critical group technique, which Dr Preston himself has developed, the critical material was the seaweed which grows between the low-tide and high-tide levels around the Cumberland coast, the seaweed used to make that Welsh delicacy laverbread. It was found that one particular radioactive product, ruthenium 106, was concentrated by the seaweed by a factor of about 100 compared with the concentration in water, and very extensive investigations were made to follow the course of this contaminated seaweed until it finally found itself in Swansea Market in the form of laverbread. I understand that there has been a television programme a short time ago showing people being questioned and radioactive counts being made of the laverbread. Of course, when you see a Geiger counter going, you immediately think there is a very large amount of activity in some small sample, but one has to realise that the detection instruments are fantastically sensitive – they are counting individual atoms and it requires very many millions of atoms to do you any harm. One must not feel unduly concerned by that. It was found that there is a critical group of about 30,000 people in South Wales who eat this laverbread. They went very carefully into the amount consumed; they found that some people had a most voracious appetite for this bread – they could swallow about a pound or so a day – and one had to see what sort of dose they were getting. Similarly, the seaweed also seems to be on the critical path from the point of view of the effects of the alpha activity. The calculations involved in this study are very detailed, too detailed to go into here, but it was clear that great precautions have been taken to limit to a harmless level the amount of effluent into the Irish Sea from the Windscale plant. This plant is, in fact, processing the greater part of the rods which come from British nuclear stations and also from nuclear stations which have been sold overseas. Indeed, the care taken and the way in which these things have been done there provide an object lesson which could well be learnt by those who are concerned in oil and chemical pollution. However, one must remember that the amount of electric power from nuclear sources is expected to increase by something up to two orders of magnitude before the turn of the century, so that the problem will become correspondingly greater.

The hope is that the purification methods, the methods of separating the highly active part from the low active waste, will also be improved in efficiency, so that there will not be anything like this large factor of increase in the effluents into the sea, in proportion to the amount of nuclear power generated. One can say that there seems very little danger indeed from the present level of radio-activity in the effluent but, of course, if this did increase by orders of magnitude, then one would have a very different story to tell. But, as I said, it seems to us that the really crucial problem in this is what happens to the highly radioactive wastes which are still being stored. We will have to find, in the course of the next twenty years, some effective way of dealing with this radioactive material.

When I say that there appears no danger at all from the radioactive effluents that are going into the sea, let us be really clear what we are saying. I have been referring to the somatic effects of radiation. Certain levels are given for the amount of radioactive material which can be deposited in the bone. For instance, the maximum body burden of an alpha-emitting isotope is put down as 0·1 micro-curies. This figure is based largely on measurements made on the skeletons of the radium painters who used to lick their paint brushes in the course of painting luminous dials. Many of them died later from the effects of bone tumours, and, in fact, of all those that have been examined, nobody who developed bone tumours had a burden of ingested radium in the bone of less than

0·5 micro-curies, and most of them had much more, so that one could say that the level of 0·1 micro-curies – I am not going to define the micro-curie, it is a measurement of radiation – would be very unlikely to result in the development of bone tumours. This goes for all the other critical organs, which are the basis for the figures given in the tables usually employed to fix specific permissible doses. But the difficulty is that there are some effects of radiation which cannot be quantified or have not yet been quantified, and it is these effects – some of which Lord Hodson referred to – which must give one cause for concern, and must prevent one from making too-sweeping statements about things being absolutely safe.

I think most people would agree with the statement that all radiation is harmful and all radiation must be reduced to the lowest possible level.

What other effects are there? Well, everybody knows of the genetic hazard, of the mutations which can be produced by the effects of radiation and the fact that most of these mutations are harmful mutations. Now, we are all subject to mutations by the effect of the background radiation. There may be a very large number of mutations, but if the effect of some increased source of radiation is only increasing the overall effect of a small fraction of the background radiation, then one has to take this in perspective, so that one can use the natural background radiation as some kind of measure of what one can tolerate from this point of view. Also, in the case of genetic hazards, it is a question of what fraction of the population is affected. In the case of the laverbread eaters, there are 30,000 of them in a population of 50 million, and so the overall genetic effect on the whole population is reduced by a large factor, because of the small size of the sample exposed. But, of course, there are other unmeasured hazards resulting from radiation. There is the effect on the life-span; there are the effects on premature births; on people being prone to develop certain illnesses through exposure to radiation.

You may recall that only a few months ago we had in this country Professor Sternglass, from the United States, who drew alarming conclusions from an analysis of the number of children dying in early years and the number of still-born children following the radioactive fall-out after certain tests in Nevada. He claimed to have found a definite correlation and, if this were true, it would be very alarming indeed. I personally am sceptical about these conclusions, but I heard a television programme in which Professor Sternglass was opposed by one of his chief critics, Professor Rotblat, with Professor Maynard Smith more or less holding the balance between the two, and I came away with the impression that the case was not proven. I could not accept all Professor Sternglass had said. On the other hand, it seemed to me that sufficient doubt about these effects of radiation remained to make us cautious before giving a completely clean bill of health. All one can say, then, is that from the point of view of compliance with the permissible doses laid down by the International Committee on Radiological Protection, and based on the somatic effects of radiation, what has been done in the nuclear power industry is certainly very adequate. But one must always have in the back of one's mind these other uncertainties, imponderables, things we do not know enough about of some of the other effects of radiation.

I think that in all these pollution problems we have to weigh the beneficia results to the community of the particular technological development we are concerned with against the associated perils. The atomic energy industry does carry with it the dangers of a certain amount of radioactive pollution. We have the problem of accidents to nuclear power stations, which Lord Hodson has referred to, and we also have the problem of the disposal of radioactive wastes. But what is the alternative? If one uses conventional power stations, then the air is polluted with sulphur dioxide. How many people are killed each year or suffer premature death through inhaling the fumes of sulphur dioxide in the atmosphere produced by conventional power stations? I have heard it suggested it may be several hundreds. If this were correct, then on this scale of comparison the nuclear energy industry might be thought a very safe industry. I think there is a standing challenge for my engineering friends in the power industry and the chemical industry to apply equivalent standards in their approach to pollution as has been applied in the nuclear power industry.

Finally, I would just like to make a general remark about this question of pollution. You see, as a university teacher, we have a few problems with our students these days and some of the more high-minded, idealist students are very concerned with the problem of pollution, and they do, in fact, tend to regard this as a kind of critique of science and technology itself. In fact, there is an alarming tendency in some quarters to go back to the sort of Luddite idea that, "Look at the dreadful mess that science and technology has brought to civilisation". It tends to bring an anti-technological, anti-scientific approach, because we are blamed for the evil effects of pollution. If science and technology can develop processes that produce pollution, then a further effort, using the methods of science and technology, can work out ways of dealing with that pollution. I think it is very important that we always bear in mind that science and technology are a great progressive force and that the problems of pollution, however complex, are capable of rational solution. They must not be used as an excuse to oppose further technology and further development of science. In fact, they are rather a challenge to us. The real crux of the matter with regard to the problems of pollution comes down to the question of how much money is spent on their solution. Certainly these questions are capable of technological solution, but at a price. We have to get our industry to understand that this price must be paid. I think that is the important lesson which a Conference such as this could help to put across. If it must be paid, then, of course, it must be paid by all industry and all countries, and therefore I think it is very important that international law is so strongly represented here, because in the final analysis one can only solve these problems of pollution, whether by chemical or radioactive waste, by international agreement. In conclusion I would like to put it to you once again that, in relation to international agreements also, those working in other fields could well learn some lessons from the way the problems have been tackled in the case of radioactive pollution.

Nuclear and Thermal Waste Pollution

MR A. PRESTON (Fisheries Laboratory, Lowestoft)

I am a scientist despite the designation "administrator" given on the programme and, in a way, I am here by default because it was not possible to provide one of my administrative colleagues for this particular weekend.

I want, first of all, to say something which follows on from what Professor Burhop has just been talking to us about. The basic standards for the control of radioactive waste disposal stem from the recommendations of the International Commission on Radiological Protection. These are recommendations which find a very wide acceptance throughout the world, including those countries which we have, perhaps, almost traditionally come to expect not to accept such things – I refer, of course, to the Eastern bloc countries – but, even there, ICRP and its recommendations are generally accepted and well regarded. Their application to control of effluent disposal does vary from country to country, as it must fit the particular administrative and legislative framework of the country concerned.

In terms of waste disposal, and particularly in terms of waste disposal into the marine and fresh water environments, guidance as to the way in which the recommendations should be applied has been given, not only by the Commission itself but in various documents produced by the International Atomic Energy Agency, a United Nations organisation housed in Vienna, which may loosely be called codes of practice. The most famous of these is probably the one published by the IAEA in *Safety Series No. 5* which is called "The Control of Radioactive Waste Disposal to the Sea", and has come to be known as the Brynielsson Report after the name of the chairman. This lays down a method by which the recommendations of the International Commission can be applied to radioactive waste disposal situations in a marine context.

In the United Kingdom we follow a pattern, and have followed a pattern for many years, close to that recommended by the Brynielsson Report. The basic recommendations of the ICRP are first considered in this country by the Medical Research Council and, in particular, by its Committee on Protection from Ionising Radiation. After consideration, and endorsement where appropriate, and within the field of radioactive waste disposal this endorsement has always been quite complete, the recommendations of the Commission are then built into the policy for the control of radioactive waste disposal.

In 1948 Parliament passed a piece of legislation called the Radioactive Substances Act which, amongst other things, set up a committee called the Radioactive Substances Advisory Committee whose principal function it was to advise Ministers in relation to their responsibility for control of radioactivity in various contexts, not only that of waste disposal. This Committee set up a sub-group to consider problems of waste disposal, and it reported to Parliament in 1959 in the form of a White Paper (Command 884), *The Control of Radioactive Waste Disposal*. This White Paper reviewed the practices of the past several years and took a glimpse into the future and made certain recommendations which, it said, should form the basis for the necessary legislation. It set three objectives for the control of radioactive waste disposal. The first two of these objectives are mandatory and have to be observed whatever the cost. The first is that "No person

19

in the United Kingdom shall be exposed to levels of radiation during his lifetime greater than those recommended as safe dose limits for members of the public by the International Commission." Secondly, "The population as a whole in the United Kingdom shall not be exposed to more than a given level of radiation – 1 rem per 30 years". This is a genetic consideration whose objective is to limit the radiation dose to the germ plasm pool of the nation as a whole, so as to maintain possible genetic damage at an acceptable level. The third objective laid down for the control of radioactive waste disposal was to work as far below these upper and obligatory thresholds as possible, and to require adequate justification on the part of an operator of the need to dispose of radioactive waste.

These objectives were given teeth in legislation at a number of points, but the principal one was the Radioactive Substances Act of 1960, which subjected all radioactive waste disposal to authorisation, but in the context in which we are talking this morning, that is mainly of the major disposals of radioactive waste, I want to confine myself to the disposal of waste from the major installations, that is, the nuclear power stations and the sites of the Atomic Energy Authority, Ministry of Defence (Navy) and other such major users. Here within England and Wales all disposals of radioactive wastes are subject to authorisation, and the authorisation is issued jointly by two government departments, that of Housing and Local Government and my own Ministry, Agriculture, Fisheries and Food. In Scotland similar responsibilities are placed on the Secretary of State, and in Northern Ireland on the Department of Commerce. In fact, in terms of disposal of radioactive waste into water, the effective scientific guidance and advice is provided by Agriculture, Fisheries and Food to the Scottish authorities and those in Northern Ireland when the need arises. These authorisations specify the safe amounts of radioactivity which may be introduced into a given environment and take note, as it were, of the need of the operator to discharge radioactive waste. They are never issued until a very careful assessment has been carried out as to the consequences of releasing the radioactive material – the critical path assessment to which Professor Burhop referred. It has as its objective the isolation, identification and quantification of those pathways which will lead to the greatest rate of human radiation exposure as a result of the intended waste disposal operation. It takes into account in a step-wise sequence the effect of dilution after introducing the radioactive material to the water, its re-concentration in biological and physical materials, the uses of these biological and physical materials which may expose members of the public, and then, against these uses, re-concentrations and the levels recommended by the Commission, maximum rates of discharge are calculated. Because of the nature of the calculations there are reservations, of course, in this type of calculation, a deliberate safety factor is introduced and, during the provisional period of an initial authorisation, the discharge of radioactive material is not permitted at levels greater than one-tenth of those calculated to be safe.

After a period of discharge and the establishment of a relationship between the amounts discharged and the levels of radioactivity measured in the environment, it is possible, if the operator can show a real need, to revise the authorisation on actual operational data. The authorisation also provides the Department with an opportunity to specify monitoring requirements, and the operator is obliged to sample, in a way specified by the authorising department, and to measure the composition of his effluents. He must use sampling and analytical methods which are either recommended by the departments or which, if suggested by the operator, are approved by the departments. He must also sample the environment in a way the departments think is adequate to demonstrate the acceptability of the discharge. He will be told what materials to sample, where to sample them, with what frequency he must sample them, how to sample them, what to analyse for, what methods of analysis to employ, and to whom and when to report the results. In addition, if the discharge is of a major character, independent monitoring will be carried out by the authorising department, so as to ensure that the procedures have been complied with. All establishments will be regularly visited, the major ones at least twice a year, to examine procedures, discuss problems, and try to develop common approaches to new problems as they arise – such a problem as the one Professor Burhop referred to of an increase in the alpha radioactive discharge at Windscale.

There is, therefore, very careful control exercised throughout the United Kingdom on radioactive wastes disposal.

There is one further point I should like to emphasise. As Professor Burhop rightly said, the vast majority of radioactive material arising from peaceful use of nuclear energy in the United Kingdom is never introduced into the environment – some 99·9 per cent is stored. It is only those very low level and very large volume streams of radioactivity, which economically it is just not justifiable to treat, which are introduced into the environment, where, under the procedures of control that I have outlined, they can be suitably diluted and dispersed without any adverse effect on the environment. There are problems, of course, to be resolved in the future, not least that of the storage of high-level radioactive waste, which is adequately contained at the moment in the storage facilities referred to. A great deal of experimental work has been carried out, both in this country and in the United States, on other ways of treating and storing high-level radioactive waste, probably in the form of ceramic materials of very low leachability which might possibly, under suitable conditions, be introduced into the deep sea in designated disposal areas where their very low leachability would ensure only very low concentrations of radioactivity in the water. Because of the isolation involved in the very great depths of the ocean it would preclude any possibility of the material coming back into contact with man on a time-scale which was unacceptable in terms of the radioactive half-life of the material concerned.

There are a number of other points arising from remarks in Professor Burhop's paper and in the introductory address by the Chairman, but I prefer now, having outlined the basis of control, to stop and to deal with factual matters as they come up in discussion.

Nuclear and Thermal Waste Pollution: Some Legal Considerations

Dr D. W. Bowett (President of Queens' College, Cambridge)

1 The Nature of the Problem

Whilst States can, and do, make arrangements for waste disposal within their own territory there are many circumstances in which either the disposal takes place beyond the national jurisdiction or the effects of the disposal are felt beyond the jurisdiction: at that point the problem must be regarded as an international and not a purely domestic problem.

The high seas are used for dumping of waste derived from shore establishments;[1] nuclear-powered vessels may discharge low-level effluents,[2] or, following a collision, release high-level radiation; vessels carrying cargoes of nuclear materials may suffer similar collisions or be wrecked; nuclear testing may occur on or over the high seas; and, in the future, the sea-bed may be used as the site for nuclear power plants.[3]

Waste may also be discharged directly into the coastal or territorial waters of a state, usually by pipeline, but thereafter be carried either by the current or tide or even living organisms into the high seas or an adjoining state's national waters.[4] So also may waste discharged into national rivers eventually reach the sea. And, obviously, waste discharged into an international river, or into any shared water like the Great Lakes, may affect other riparian states.

Finally, gaseous wastes released into the atmosphere or fall-out from nuclear testing may affect either the high seas or other states.

Given the variety of sources of nuclear and thermal waste, the different means

---

[1] Burns, "Radioactive Waste Control at the U.K. Atomic Energy Research Establishment, Harwell" in *Disposal of Radioactive Wastes*, IAEA publication (1960), Vol. 1, 413 at 426–7. See also the series of papers of the *Vienna Symposium on disposal of Radioactive Wastes into Seas, Oceans and Surface Waters*, IAEA publication (1961). In 1967 a group of European States, in an undertaking sponsored by ENEA, disposed of 11,000 metric tons of radioactive waste into the Atlantic.

[2] Pritchard, "The application of existing oceanographic knowledge to the problem of radioactive waste disposal into the sea" in *Disposal of Radioactive Wastes, op. cit.*, Vol. 2, 231 at 243–7.

[3] Suggestions have been made for such power plants to provide electricity for underwater cities (Burgio, "Radioisotopes in the Marine Environment", *The Decade Ahead, 1970–1980*, the Marine Technology Society, 1969, p. 153) or to provide power for deep sea-bed drilling (Edwards and Zupanick, "Floating Powerplant to Support Submerged Offshore Operations", First Annual Offshore Technology Conference, 1969, Paper No. 1131, Vol. II, 481).

[4] Many instances are given in the *Vienna Symposium, etc.* (footnote 1 above). Barnes and Gross (*ibid.*, 291) reveal that low concentrations of radionuclides from the cooling water of the Hanford reactor in the Colombia River are carried 650 km. into the ocean; Carey, Pearcy and Osterberg (*ibid.*, 303) show that Zinc-65 from this same reactor is detected in fish as far as 490 km. off the coast of central Oregon. Feldt (*ibid.*, 739) shows the progressive increase over six years of radionuclides ^{90}Sr and ^{137}Cs in fish, mussels and shrimps in the North Sea. One of the most disturbing discoveries in recent years is the capacity of some marine organisms to accumulate toxicity, thus triggering off a toxicity cycle with the result that initially harmless levels of toxicity become concentrated in the organism and ultimately harmful. Thus dispersal of nuclear wastes is not the inevitable result of disposal in the sea.

of transmission and different levels of activity,[1] there is here a complex problem. There can be no easy solution in terms of a total ban of waste disposal beyond national limits: such a solution is not only impractical but also ignores the very considerable potential of the high seas as a safe disposal area provided the disposal is regulated and controlled. There will thus have to be a regulated use of the marine environment as a disposal area with the regulations striking an acceptable balance between the nuclear energy users – or operators – and the competing interests of navigation, fishing, coastal amenities and preservation of marine life and fauna. To strike this balance will be no mean task. What is "safe" is essentially a scientific question, and it may well be that the data essential to answer this question is not in every case available. But the further question of what should be permitted is a separate policy question, by no means coincident with the first, and this depends upon the weighing of values.

2 Existing Regulation

At the national level, certainly in states like the UK,[2] the USA[3] and France, legal and administrative control over all nuclear activities, including waste disposal, is very strict. However, once the problem is conceded to have international implications – as must be conceded – it becomes clear that regulation by national legislation cannot suffice. Not only is there the risk that some states may, in their national legislation, fall short of the standards objectively required by the international community interests but also there is the risk that the cumulative effect of the disposal activities of several states may create a hazard of which they may be completely unaware. Hence, some form of international regulation and control is, in principle, essential to a solution of the problem.

In fact, however, regulation and control at the international level is minimal. There exists only the broadest, and vaguest obligation in Article 25 of the 1958 High Seas Convention:

"1. Every state shall take measures to prevent pollution of the seas from the dumping of radioactive waste, taking into account any standards and regulations which may be formulated by the competent international organisations.

2. All states shall cooperate with the competent international organisations in taking measures for the prevention of pollution of the seas or airspace

[1] Most waste disposal into the marine environment is believed to be of low-level or, at most, intermediate level activity. However, studies are being made of the possibility of storing high activity wastes in solid form in the ocean depths: see Dunster and Wix, "The Practice of Waste Disposal in the U.K. Atomic Energy Authority" in *Disposal of Radioactive Wastes*, Vol. 2, 405.

[2] The basic legislation is the Atomic Energy Act, 1946; Radioactive Substances Act, 1948; Atomic Energy Authority Act, 1954; Radioactive Substances Act, 1960 and Nuclear Installations Act, 1965. A good commentary on the UK legislation can be found in Street and Frame, *Law Relating to Nuclear Energy* (1966).

[3] In the USA control is both federal and state. Apart from the federal Atomic Energy Act of 1954 and the supervisory control of the Atomic Energy Commission, there is a great deal of state legislation and control of disposal of wastes, pollution, transportation of radioactive materials: see Stason, Estep and Pierce, *Atoms and the Law* (1959). For a somewhat alarmist view of the efficacy of these controls see Curtis and Hogan, *Perils of the Peaceful Atom* (1970).

above, resulting from any activities with radioactive materials or other harmful agents."

There is, of course, the 1963 Test Ban Treaty,[1] which is important because, to date, nuclear tests have been the largest cause of radioactivity in the seas.[2] But China and France are not parties to this Treaty. Nor is the joint US/USSR draft treaty for the prohibition of the emplacement of nuclear weapons on the sea-bed, submitted to the General Conference of the Committee on Disarmament, likely to achieve universal acceptance.

There are, in addition, the IAEA regulations for the safe transport of radioactive materials[3] and a separate set of regulations governing disposal of waste; but the former constitute a code for the guidance of national legislation, not in itself binding, and the latter bind only states accepting fissionable material from the IAEA.

This body of international regulations is clearly inadequate to the task. It is therefore necessary to consider what could be done in the future.

3 A desirable Regulatory and Organisational Structure for the Future

(a) *Prevention of hazards*

There is a clear need for a comprehensive international convention setting out the specific measures which states should take to avoid pollution. This is now increasingly recognised. The General Assembly, in calling the 1972 UN Conference on the Human Environment, endorsed the idea that the foundation of new controls must lie in international conventions. The IMCO Conference planned for 1973 on marine pollution has the express aim of preparing international conventions. Thus, the opportunity will soon present itself for initiatives to be taken in drawing up suitable international conventions. Quite how specific and detailed these could be is a matter on which scientists may have a clearer view than lawyers, for much depends upon the accuracy of the assumptions made about the levels at which a hazard is created, about absorption capabilities of the sea, tidal movements, etc. However, the IAEA has already formed a series of conclusions[4] which might be capable of incorporation in a convention of which the principal ones were (i) high-level wastes from irradiated fuel ought *not* to be disposed of at sea, (ii) the dumping of low or intermediate waste should be controlled and confined to selected disposal sites (except for nuclear ship effluents), (iii) all disposal should be recorded on registers, and (iv) the IAEA should maintain a central register of disposals and of all monitoring programmes – these to be carried out under techniques standardised by the IAEA. Again, I would stress that the need for a continual system of monitoring seems generally recognised. This was the firm conclusion of the International Oceanographic

[1] *U.N.T.S.*, Vol. 480, p. 43.
[2] See statement by IAEA in "International Co-operation in Problems related to the Oceans", progress report, UN Doc. E/4911, 28 May 1970, para. 31.
[3] "Regulations for the Safe Transport of Radioactive Materials, 1964 Revised Edition", *Safety Series No. 6*, IAEA, Vienna (1965).
[4] *Radioactive Waste Disposal into the Sea*, IAEA Safety Series No. 5, Vienna (1961), Conclusions.

Commission (IOC) in 1969, of GESAMP,[1] of the FAO[2] and many other bodies.

It may well be that a general convention could be far more detailed than this: this would depend upon how great a measure of consensus on the specific and other issues could be reached. Certainly one is encouraged to see how detailed the IAEA Regulations for the Safe Transport of Radioactive Material are. But we may have to be content with an initial Convention confined to broad principles which could be supplemented by Annexes or more detailed regulations as scientific experience and a general willingness to co-operate grows.

Such a general Convention would, in scope, cover the high seas and, I would hope, territorial waters and even, possibly, national waters which are tidal or which ultimately flow into the sea.

I would not regard such a general Convention as appropriate for international rivers or inland waterways like the Great Lakes: these I would expect to see governed by separate, special conventions between the riparian states in far greater detail than any general Convention.

Conventions are not, in themselves, enough. Agreed standards would require constant supervision in the form both of monitoring and of scrutiny and assessment of the reports derived from monitoring: hence some organisational machinery is necessary.

Doubtless in many areas the task of monitoring could be left in the hands of national authorities. But it is difficult to see how some form of international monitoring, by an international, scientific inspectorate, could be avoided. Not only will it be important for states to have an independent check on reports from national authorities and an assurance that the standardised monitoring techniques are adhered to, but there may be areas of the world in which national states are simply unable to assume the burden of necessary monitoring: the Indian Ocean is possibly such an area. I would also foresee a need for review organs of the international community at two distinct levels: the first, expert, scientific assessment and analysis of all the reports stemming from national and international monitoring systems; the second a more political level of review at which state policies could be called in question. There is no necessary need to think in terms of supranational bodies. Organisations like OECD, GATT, the ILO and EFTA manage to exert a good deal of pressure on states through a process of "confrontation" in which states are confronted by evidence of non-compliance and required to defend their conduct. This seems exactly the kind of process best suited to the problem of securing compliance with agreed standards on pollution.

Apart from these supervisory functions at both the scientific and political levels, there is a case for attributing to an international agency two additional functions.

The first is a legislative or quasi-legislative function. Time and experience will surely reveal the need for additional Conventions or supplementary Protocols and the initial drafting and discussion of these would rightly be considered a

---

[1] Group of Experts on the Scientific Aspects of Marine Pollution, a joint IMCO/FAO/UNESCO/WMO/WHO/IAEA body: see GESAMP II/W.P.1 Rev. 1, March 1970, para. 5.

[2] The FAO propose to hold in December 1970 a Conference on Marine Pollution, mainly because of their concern to preserve fish stocks, and will discuss the possible establishment of an international toxicological surveillance system.

proper function for the competent international agency. These new Conventions would not be imposed upon states but would become binding through the normal process of ratification. In addition, a more truly legislative function might be conferred in relation to Conventions already in force; that is to say, assuming a basic Convention, this might be supplemented by more detailed regulations on matters such as safety precautions, packaging of disposal containers, etc., where the basic Convention contains the broad agreement of principle but detailed regulations can be worked out by an international body and, upon promulgation, become automatically binding on all states.[1]

The second additional function is disputes settlement. Disputes arising from pollution by radioactive agents are likely to depend upon highly technical, scientific forms of proof and may well require an expert body quite different from the ICJ or the usual arbitral body composed of lawyers. Again there is no particular novelty in the notion of a body with arbitral functions being composed of scientific experts; the concept has been used in fishery disputes for years. The principal difficulty will be to persuade states to accept a compulsory jurisdiction for such a body. Indeed, it may well be that, if the organ of political review can develop the confrontation techniques to which I have already referred, and can utilise advisory panels of experts to assist it, then we can dispense with formal disputes machinery entirely. This would avoid antagonising those states which object to compulsory arbitration.

A final, and very difficult question is which international agency ought to assume these various functions of research, monitoring, political and scientific review, initiating new or supplementary legislation and, possibly, disputes settlement? There are those who advocate an overall approach to the whole question of preserving the environment and therefore propose *a new International Environment Agency*. This I would regard as over ambitious. I would share the view that *all* problems of marine pollution and environmental protection ought to be covered by one body, if only for reasons of economy and in order to avoid duplication of research and monitoring activities. Already, as GESAMP demonstrates, IMCO, FAO, UNESCO and WMO have an interest and the IAEA is another obvious candidate. No one of these existing agencies suffices for an overall control over the marine environment and there would therefore seem a case for a joint venture to which they all subscribe, which would have a quasi-autonomous status but be responsible ultimately to them all. There is something of a precedent for this in the World Food Programme, operated jointly by the UN and FAO. This organisational problem is crucial, for unless it is solved I can see all the good intentions being brought to nought.

(b) *Liability for damage caused*

Liability presupposes proof of causation and this is primarily a scientific matter. Assuming that evidence of a causal link between damage and the acts of either a

---

[1] This is not particularly novel. In the WHO the Assembly adopts regulations on matters like sanitary and quarantine procedures, nomenclatures for diseases and pharmaceutical products which are binding on members unless they expressly opt out. In ICAO the Council adopts international standards to supplement the ICAO Convention in the form of Annexes to the Convention and these become effective within three months unless a majority of members register their disapproval.

state or non-state activity can be established, we then have an issue of legal liability. I do not wish to expand on this overmuch, because prevention is much the more serious problem.

Within municipal systems the pattern of legislation now emerging is to channel liability on to the operator of a nuclear installation, to make the liability well-nigh absolute, to require any non-state entity to insure against liability and to have the State ready to indemnify – as a kind of guarantor – up to defined financial limits. Under the 1954 US Atomic Energy Act this limit was $500 million per accident. Under the UK Nuclear Installations Act of 1965 the licensee, operating a nuclear plant, insures or otherwise provides cover up to £5 million, and claims in excess of that figure are made to the Government.

The system at international law is not markedly dissimilar under the three Conventions now introduced (although these are limited in the sense that they bind only contracting states). The 1960 Paris Convention on Third Party Liability, as supplemented by the 1963 Paris Convention, channels liability on to the operator within maximum financial limits. The Vienna Convention of 1963 follows similar principles, with the operator being required to provide cover up to £5 million and the Installation State standing as guarantor. The Vienna Convention, negotiated under the IAEA, aimed at embracing many more states than the basically European Paris Convention. There is a separate 1962 Brussels Convention on the Liability of Operators of Nuclear Ships, following similar principles but with a limit to liability of $100 million.

I would not expect this general pattern to change. What is now required is the encouragement of all states to participate in the IAEA Vienna Convention, and the Brussels Convention, with some thought being given to an international fund to ensure payment of claims when either the operator or the Installation State or Flag State of the vessels fails to do so.

Discussion

DR J. A. G. TAYLOR (Unilever Research)
Could I ask Dr Preston how well he thinks he can identify this critical pattern of disposal? The situation will be a very complex one and we can only look at the short-term effects.

MR A. PRESTON
I am not claiming that it is easy. It requires a year or two of work at any particular site to do this. We have experience in the United Kingdom, I suppose, of a couple of dozen sites: nuclear power stations, Atomic Energy Authority sites, and so on – and we have developed some very interesting approaches and some sort of feel for the subject – but, basically, the identification of critical paths comes, first of all, from a knowledge of the radionuclides which are to be introduced into the environment and from the very considerable scientific knowledge, now, of the way in which they will behave in terms of their partition between water and sediment, the degree to which they will re-concentrate in biological systems, and so on, and from this knowledge it is possible, even sitting down at a table with a piece of paper, to suggest, in a given situation in UK coastal waters,

where the radioactivity is likely to end up, in terms of potential radiation exposure.

DR J. A. G. TAYLOR
What sort of mass balance can be achieved?

MR A. PRESTON
In the Irish Sea or from Windscale? Not bad. In some situations, not at all good because the amount of radioactivity being introduced is so small that a positive and accurate measurement in any given segment of the environment is not possible. But where it is possible, pretty good estimates can be drawn up of compartment sizes.

We then proceed to what we call a "local habit survey", and I use the word "local" advisedly because, although it starts in a "local" environment, it clearly, at times, gets much farther afield, e.g. to South Wales, and may even get to London's oyster bars and various other interesting places. From this habit survey we try to make an assessment, quantitatively, of the amounts of material eaten by people making due allowance for wastage and processing and so on. We make estimates of the time spent on beaches, for carrying out commercial fishing operations, or digging bait, all the hours spent by dredgermen dredging rivers, and things of this kind, and from this sort of data, build up over several surveys, covering a period of months and contrasting seasons, a picture of the hours spent on beaches, the so-called occupancy factors – in the jargon of the health physicist – and the rates of consumption of contaminated materials.

In a very comprehensive survey of this type carried out at a nuclear power station, you can literally physically identify the people concerned, and it is possible, of course, to be quite accurate and quite comprehensive in the study of consumption, etc. If, however, you are dealing with a population as big as that in South Wales, it clearly is not possible to identify all concerned and to quantify their habits. All you can do is to try to get a representative sample of the population and, from the distribution of habits in this sample, extrapolate to and infer something as to the habits of the population as a whole.

If you look at the consumption habits, for example, of the South Wales laverbread-eating population, which is a little over 26,000 people, we took a sample of something like 2,000 of them, which is a pretty big sample, selected randomly, and the data was obtained by interview on the principal market days in the Welsh cities, towns and valleys. Customers were interviewed at the time of their laverbread purchases about the eating habits of their families, and, when this data was analysed, it fitted very well into the log of normal distribution, except for a very small group of consumers, some 7 in the sample, and some 100 in the population as a whole, who had exceptionally high rates of consumption, so high that you could not have forecast them from the behaviour of the rest of the distribution in the sample. These were taken to be the critical group of the population, and the control at Windscale is set so that things are absolutely safe for this exceptional group.

In a power station situation, where we can be more positive about the identification of people and their habits, we again set out to cater for the exceptional individuals and not the average in the population. In most cases, in fact in

28

all cases in the UK, the authorisations that are issued, and the levels at which these are operated, in terms of the amount of radioactivity actually being discharged, are well below that considered to be safe. We are, in fact, only using a small fraction of the acceptable rate of exposure even for the exceptional people.

Dr G. Brown
I would like to suggest that there is another kind of exposure that has to be taken into consideration, and that is the exposure of the human race to X-radiation which, I believe, may well be considerably higher, on the average, than the exposure of the human race to radiation from radioactive sources. Could anyone give us definite figures on this?

Mr A. Preston
I think we should get Dr Marley to answer this; it is one in a broader context than mere waste disposal; it is the exposure of the population at large. What is specifically being referred to here is the use of X-radiation in medical practice.

Dr W. G. Marley
This is correct; that exposure in the highly developed countries to analysing radiation of man-made origin comes mainly from medical practice. This subject has been reviewed by the United Nations Scientific Committee on the Effects of Atomic Radiation, and figures for the level of exposure in several countries from data submitted were given in their report issued in 1962. The UK situation was explored under the aegis of the Adrian Committee and the figures, if I remember correctly, were that the genetically significant dose to the UK population was about one-fifth of the exposure due to natural radiation. The important dose is, of course, diagnostic radiation, because this may carry genetic effects, whereas therapeutic radiation is unlikely to do so.

Now this subject is, of course, being very carefully considered by those responsible for the control of medical practice and procedure. I think the position has improved considerably since this enquiry was made a few years ago. The position in America is very much worse than in the United Kingdom because of the very wide use of X-rays by general practitioners.

Dr W. R. P. Bourne
There are two points I would like to make. One is that I heard that early last year there were complaints from the Russians that the Irish Sea was becoming radioactive. I never understood whether there was any real substance in this, so I wonder if we could be enlightened about it. The second thing that I am somewhat curious about is whether the radioactive substances that are released are fairly uniform in composition, or whether they are a mixture, and, if so, whether there might be some important trace elements among them whose fate would also be interesting – substances like copper and cobalt are very important in very small quantities.

Mr A. Preston
By virtue of the Windscale discharges, the Irish Sea is more radioactive than it would otherwise be, but the levels of radioactivity, as I think I indicated in my talk, are determined – and certainly Professor Burhop referred to this – by the

29

safe levels in seaweed on the littoral margins of the Irish Sea, particularly those of Cumberland and the north-west coast of Lancashire, where the highest levels of radioactivity are found. This is so simply because these are the areas closest to the point of introduction of radioactive material. As one moves away from the point of introduction towards either the coast of Northern Ireland or north or south, there is a very rapid fall-off in concentration of radioactivity, both in the water and on the sea-bed. What happens in practice is that when the radioactivity is introduced, and it is introduced to the sea via a pipeline one and a half miles long laid on the sea-bed and the effluent is somewhat less dense than sea water, it rises vertically after release and as it rises it is sheered by the movement of the tidal stream so that elliptical contours of radioactivity of decreasing concentration are built up as you go outwards in any direction from the pipeline. The concentrations fall most rapidly, therefore, as you go across the Irish Sea, less rapidly as you go north and south, but the levels are everywhere quite safe and quite low and require, except in the immediate vicinity of the outfall, quite sophisticated methods to measure them. I can add to that, if you want me to, later.

Turning to the second question, of course the effluent composition at the various stations does differ somewhat. At Windscale we are dealing mainly with fission product radioactivity and the principal constituents we are concerned with – this may mean nothing to a lot of the audience – are the following radionuclides: 106 ruthenium, 144 cerium, 95 zirconium, 95 niobium, 137 caesium, and 90 strontium. Now, strontium, of course, is not a trace element in a marine context, there are about 8 parts per million of strontium in sea water, and there are at least 400 parts of calcium, which, metabolically, is a very similar element, and so there is a massive isotopic dilution of the strontium 90 in a physiological or metabolic sense. The remainder, of course, are trace elements in water, and exist in concentrations which are measured in terms of parts per thousand million or less. However, none of them are trace elements of metabolic significance, that is they have no biological significance, they are not elements which are essential metabolites in any way and so they are not substantially re-concentrated.

If we now turn to a nuclear power station, where we are not dealing, in essence anyway, with fission product radioactivity, but with radioactivity which has come about by virtue of the exposure of the cladding material in which the fuel is wrapped – it is a magnesium alloy can – to neutrons; radioactivity is induced in this can, and subsequently this radioactivity is able to get into solution when the fuel elements are withdrawn from the nuclear reactors and placed under water in large cooling ponds, where they are allowed to cool in both a thermal and a radioactive sense, for some little time before being transferred to Windscale for treatment. Here we do see isotopes like 60 cobalt, the one you mentioned, 65 zinc, 54 manganese, 55 iron, 124 antimony and a number of other metals. Many of these metals are quite important metabolically – they are essential for life. They are at trace concentrations only in the sea water and, in general, because of that, this is where we see the highest degree of re-concentration in biological material; you will find much higher concentration factors, that is the ratio between concentration in the water and concentration in a biological material, for the radionuclides for these elements, which, I think, is probably the point that Dr Bourne was leading up to.

Just to set it all in perspective again, remember that these are the very considerations that we are taking into account when we are assessing a given disposal situation. We know well the values for these concentration factors and absolute allowance is made for them in deriving the permissible levels of discharge.

Mr G. W. Hull
Following the last two speakers, I have a question of theory concerning the detoxification of radioactive substances in the living environment, through its biological processes. I wonder whether the arguments of Kervran concerning the biological transmutation of elements has been considered relevant?

Mr A. Preston
There are no biological processes which, of themselves, can render a radioactive substance non-radioactive; the decay of radioactivity is an immutable physical process which proceeds at a fixed rate. There are, however, various chemical techniques for increasing the rate of excretion of radioactive material once it is in a biological system such as the human body. There are various complexing agents which can be injected or fed to anybody so contaminated which will bind themselves to the contaminants in the body and then be excreted at a greater rate than they would otherwise be. Quite a lot of research has been put into this "flushing" from the human body, but in general, for many of the more serious kinds of radionuclide with very long half-lives in the body, it does not seem to be a technique which is likely to be very successful, certainly not, I think, for most of the actinide elements – the transuranics that Professor Burhop was talking about. You can increase the excretion of strontium and some other nuclides from the body by using substances like ethylene-diamine-tetra-cetic acid and complexing agents of that kind. Perhaps Dr Marley could add to that.

Dr W. G. Marley
I would like to add something about the elimination of strontium from the body. This material used to be thought a permanent acquisition in the body when it was built into the bone. However, it now turns out, as a result of very interesting work done by the Medical Research Council in their Radio-biological Research Unit conducted brilliantly by Dr Loutit and his colleagues, that the turn-over rate of strontium in young children is very much faster than in adults. So, our original concept that strontium in fall-out was primarily a hazard to children is possibly wrong. What it amounts to is that young children are turning over their skeletons completely in a period of some 18 months. If, as a result of fall-out contaminated milk, the bone is contaminated, and you subsequently supply clean milk or food to the child, the strontium rapidly disappears from the skeleton, because the skeleton is being rebuilt with new calcium free from strontium. This just emphasises two points: first, that knowledge of the metabolism of the radionuclides is increasing all the time, and that some of the views that we had earlier about the severity of the hazard are being modified by new biological information. I suppose this may mean an age transfer of the group most at risk.

31

Miss M. Sibthorp

I do not think everyone knows that there has been considerable encouragement over recent years for industry to export nuclear reactors. If such nuclear reactors are exported to underdeveloped countries, are there any facilities, either in the human sphere or in the ecological sphere, for the careful control of atomic wastes such as we have here?

Mr A. Preston

I think in general it is true to say that most of the nuclear operations that are conducted throughout the world are conducted in the more highly advanced, if we can use that term, countries, where the technological support is adequate to carry out the sorts of controls which I indicated earlier this morning. The methods which should be used, and the philosophy and the techniques, are, of course, all well understood, they are all well documented, they are all accessible in the scientific literature, so what places a limit on their application in the underdeveloped countries, of course, is the ability to implement thoroughly understood techniques and procedures, and this, if there is a limitation, is where it will arise. In general, however, in selling a nuclear reactor, or any other nuclear facility, to another country, part of the bargain, as it were, part of the contract is to provide training for people who are going to operate the nuclear reactor and are going to look to the control of waste products, and so on. This training is normally exported by the country selling the reactor. We train many people in this country, through the auspices of the Atomic Energy Authority and other organisations. Consultants and experts from the United Kingdom go out to the international organisations, they take part as experts on panels dealing with practice and procedures, they go out to individual countries to offer advice on particular situations, and so on. A very concerted effort is made to meet the point you raised. Whether or not it is entirely successful in every case is, to some extent, questionable.

Dr D. W. Bowett

Could I just add one thing? I think it is the case that where a country obtains fuel for its reactors from the atomic energy agency it does so on the condition that it subscribes to the agency's safeguards as to disposal of wastes.

Mr J. E. S. Fawcett, Chairman

I will now throw the session open for general discussion but, since our time is limited, I think it may be useful if as Chairman I guide it along fairly arbitrary lines.

One point, it seems to me, does emerge from what our three speakers have said and that is the question of safety levels. Remembering that this is not a wholly technical conference, it is a meeting of lawyers, administrators and scientists, I think what a lawyer instinctively asks about safety levels is, "How are these, in fact, enforced?" It seems to me, following what Dr Bowett said, there are two questions here. What criteria do these safety levels meet: are they the levels of natural radiation? What is the criterion by which something is said to be safe? Now here I think that there is some difference between what Professor Burhop and Dr Preston said. Dr Preston suggested that, at least in the radio-

32

active effluent in the sea, the levels were safe and controlled, but Professor Burhop seemed to be saying that there were certain effects of radiation that could not be quantified; and certainly from a layman's point of view, one might ask whether the genetic effects of radiation can be known until after a lapse of perhaps thirty or forty years. If there are effects of radiation, and Professor Burhop gave you examples in this area that are not quantifiable, what is the meaning of "safety levels"? Can safety levels – to use Dr Bowett's point about international standards – be defined and agreed for what I would call legislative purposes? Another question I would ask is what happens if the radioactive effluent is found to be above the level; is there a process of prosecution? What is the control nationally if the Ministry detects such an excess of the safety level? The second point is: can this be done at all internationally? I think we could also usefully ask, following Dr Bowett's suggestion about the international organisation of monitoring programmes and other techniques of control, how effective can this be, and should it be done regionally? Can one think of this globally?

SIR FREDERICK WARNER

Could I make a comment, first on the semantics, since this is an important point when we talk about levels. Levels are the most difficult things, whether we are talking about radioactive materials or any other form of polluting discharge, because the normal thing that an administrator or a legislator wants to put in is to say that a level can be defined in terms of a concentration, but where scientists and engineers are concerned the problem is that we can only define in terms of mass and then in terms of the mechanism which disperses to give what you would then call a level. We should try to get the semantic question clear first.

MR J. E. S. FAWCETT

Could you comment on the word "safe"? This is the key word.

MR A. PRESTON

If we can examine the problems you have just outlined, Mr Chairman, in two separate stages, and perhaps invite Dr Marley first of all to tell us something of the concept of ICRP and its definition of an "acceptable level" – I use the word "acceptable", not "safe", in an absolute sense – and then I will try to take over from Dr Marley and interpret this in a marine context. I hope to clarify some of the problems of defining the levels that Dr Bowett wants for legislative purposes.

DR W. G. MARLEY

Reference has been made to the standards recommended by the International Commission on Radiological Protection. This is an important body in a particular sense, in that it is a non-governmental international association of scientists. Countries do not nominate the members. The body is an offshoot of the International Congress of Radiology and was originally set up about 1928 because of the recognised dangers of work with X-rays in the early years of the century.

Of course, with the advent of atomic energy the main interest of the Commission has been standards of protection for the radioactive materials encountered in atomic energy work and now used extensively in industry and medicine.

33

An enormous amount of experimental work has accordingly been carried out to find how hazardous radioactive substances in the body are, and particularly the substance plutonium, which had not been encountered by man before and which will be the main fuel for nuclear power stations after the next decade. Recourse was had at first to the knowledge about the radium dial painters, to which Professor Burhop referred; this was a catalogue of disastrous industrial experiences in painting radium dials with luminous paint with a paint brush, during and just after the First World War, as a result of which, in the course of forty years, quite a number of persons died of cancer of the bone.

In recent years, of course, a vast amount of experimental data has been obtained from direct animal experiments. For instance, an important series of experiments is going on in Utah, where large colonies of dogs are being kept, some free from radioactivity and others with varying quantities of a number of different hazardous radioactive substances in them. This experiment will continue for twenty years; it is rather interesting that the life-span of the dog is found to be 30 per cent longer in these experiments than is normally experienced with dogs, because the conditions under which they are kept are rather like the conditions in this student hostel, they are very good.

Then we also have a lot of data on the effects of radiation from the atomic bomb survivors, and I think it is to the great credit of the Japanese that they have collaborated so intensively with the Americans in a long, penetrating study of the survivors of the explosions at Hiroshima and Nagasaki. This has enabled the effects of a sudden dose of radiation affecting the whole body to be studied, and the effects on the population on a statistical-epidemiological basis to be assessed.

Another set of data come from the work on the treatment of the disease ankylosing-spondylitis which is treated with heavy doses of radiation to relieve pain and suffering. It is found that persons so treated have a somewhat higher incidence of certain diseases which are known to be caused by radiation such as leukaemia and cancer. The statistics here give numerical data and I would emphasise that in this field (it is, perhaps, almost unique), we do have a very great deal of numerical data to relate to risk with level of exposure.

In interpreting all this information, the International Commission has had to estimate what the effects of very low doses are, from numbers which are derived from very high doses. For many of these effects of external radiation the data have necessarily been obtained at a dose of 100 rad or upwards, whereas in occupational exposure we want to know the effects of doses of a few rads, and in population exposure the dose of a few millirads, or thousandths of a rad. So there is a necessity to extrapolate downwards. The Commission has chosen the cautious procedure of extrapolating downwards on a linear basis. That is to say it has assumed there is no threshold for the effects. (Dr Marley illustrated this with a diagram on the blackboard relating "dose" with "effect".) To find out what the effect is at low levels of dose the curve has been drawn back through zero. The reason why the recommended permissible dose is not zero, from this curve, is that, first of all, there is natural background radiation. We are all subjected to radiation from the potassium in our bodies, uranium and so on in the walls of the building, cosmic radiation coming in from outer space, and this

34

means that you cannot get down below a certain level of dose which, according to this theory, would produce a certain effect, but the effect here is swamped in any case, because the diseases we are looking at occur in nature from other causes to a level far above what could be attributable to the radiation background. Thus you can say that, so long as the expectation of life is not appreciably affected, any induced effect from the radiation could be accepted if there is economic or other advantage to the country to incur the radiation dose. The Commission has emphasised the need to balance the risks against the benefits. For instance, the use of X-rays in hospitals is quite justified because there is great benefit not only to the people exposed but also to all of us who might require this treatment later, so we say that a certain level of dose is justifiable. It may possibly induce a small effect on the incidence of certain diseases, but so long as that effect is small compared with the natural incidence, it should be acceptable.

This has a bearing on the definition of what is "safe". I think that the word "safe" is a word which requires very great care in usage. Is it "safe" to drive your car from here to wherever you live? "Safe?" Well, it incurs a certain risk. In this case we would say, "Yes, it is safe, if you take care", but it is not "safe" with a zero risk, so we generally interpret the word "safe" as meaning "incurring an acceptably low level of risk for the benefit incurred in the procedure". In the case of atomic energy it is believed that it has benefits to confer, with progressively lower costs for power, and this is of enormous benefit, not only industrially, but also, for instance, in the public health field. It may even make a greater margin of wealth for spending on direct health facilities.

It can also be maintained, and one or two speakers, Professor Burhop, for instance, referred to the possibility, that it may be safer, in the public health sense, than other methods of generating power. If this is real, then perhaps we should be attracted to this source of power for that reason, and the economic and social benefits of a cheap source of power must be taken into the balance.

There is reason to believe, from biological experiments, that the linear extrapolation to low doses is an extremely safe procedure. We know that the rate of exposure, that is the dose rate, the rate of delivery of energy in the cells, in occupational exposure and in the case of the exposure of the public to radioactive substances, is perhaps a million times – in some instances a thousand million times – lower than the rate of exposure in the instances considered in getting this reference point on which the judgment was based.

Reference has been made, I think by Professor Burhop, to the question of the importance of being concerned about genetic effects. Twenty years ago, when many of the levels of exposure which are recommended by the International Commission were fixed, it was thought that any radiation dose had a proportionate genetic effect on the germ pool of the exposed population. This was the view of Mueller, who was a Nobel Laureate in this field, but further work has been done in the last twenty years and experiments with an enormous number of mice has now been carried out. They are called mega-mouse experiments – a million mice in each of the experiments. They are being carried out in the United States, as you might suppose. The results of these experiments now show that the concept that Mueller postulated twenty years ago is no longer fully justified.

35

The work is being done at Oakridge by a world-famous geneticist, Dr Russell, and he has made experiments progressively reducing the dose to mice populations and then breeding them, and observing the effect of the radiation in producing certain specific identifiable malformations in the progeny which can be quantified. His conclusion is that: first, there is no threshold dose rate in males, but in females there is, for all practical purposes, a threshold dose rate. Moreover he finds that the effects of radiation, particularly in females, tend to disappear if there is an interval between irradiation of the body and breeding of the female mouse. So, astonishingly, there is a healing process going on, apparently. It is only "apparent" because it is a kind of selection process in which cells with damaged genes are less preferable than cells that are undamaged by the radiation. Russell now says that, quantitatively, his experiments show that the best estimate of genetic risks from exposures of both sexes to chronic radiation is only about one-sixth of that which was estimated when the current permissible levels of exposure of the population were chosen by the ICRP. In other words, on somatic effects and on genetic effects, we are gradually getting evidence that the real dose rate dependence is more likely a curve. However, the present levels are fixed on the basis of a linear dose-effect relationship.

In this way, I feel – and my job is protection against radiation – that the present levels have a useful built-in safety factor.

Mr J. E. S. Fawcett, Chairman
I would like to ask you one question about what you have said on the mega-mouse experiments. I have seen it said, with what truth I do not know, that the acceptability of penicillin, because for example guinea pigs and rats vary very greatly in their reaction to it, might well have depended on which animal you experimented on; and all I ask is whether, in the case of the mega-mouse experiment, the extrapolation from the genetic effects on mice to the human is being clearly shown. Is there no basic uncertainty here?

Dr W. G. Marley
This is, of course, an important question, and I would like to make it clear that I am not a geneticist. I would therefore rather hesitate to say on what grounds the persons considering these genetic effects are prepared to trust the mice experiments. Ideally you should do the experiments on the species about which you require the information, but here there are practical difficulties, shall we say! A great deal of work has been done on repair mechanisms in the various species, and what is said – and I think this answers your question – is that the difference between mice and men, in this regard, is very much smaller than the difference between the fruit fly, *drosophila*, on which Mueller was basing his work, and man. The genetic constitution of the mouse is sufficiently close to that of man to suspect that the same principles will apply, though there may be quantitative differences.

Professor E. H. S. Burhop
Could we have a comment on the life-span, on the specific effects on ill-health and the incidence of disease in young children at certain stages, and on the conclusion we have drawn from Professor Sternglass's contention that there is a great deal of uncertainty about some aspects of the real situation?

Dr W. G. Marley

A word about Professor Sternglass first of all. I must say that I was very adversely affected regarding his presentation on television, when the journal *Nature* pointed out that the curve of infant mortality against data that he used was not correct and, moreover, that it was incorrect in the direction which more strongly supported his case. Now, I think this, in a scientist, is an extremely irresponsible thing to do. Even if it occurred by accident, he should have guarded against it, especially when using it to instruct an audience of 10 million people. Therefore, from the point of view of acceptability, I find it very difficult to believe other things that he says.

Professor E. H. S. Burhop

Yes, but he had some evidence from specific states in the United States. However, I quite agree with you that it is an unpardonable thing to do.

Dr W. G. Marley

May I say something more on accuracy of data? I think, in this field, that it is exceedingly important that those of us who use numbers should make sure that the numbers are right, because they get quoted, and sometimes quoted again and again. So I really must pick up a statement made by Lord Hodson because I am afraid he must have quoted from an unreliable source when he said that the amount of radioactivity released in a nuclear accident at Windscale, in 1957, was greater than that released at Hiroshima. It is ten thousand times wrong. I suspect that the statement came from a book which is full of mis-statements of this kind. Moreover, in the book, that statement is attributed to Sir John Cockcroft. Now I know for certain that he did not make it. I was working with him at that time. It was made by a newspaper reporter doing his own calculation on something Sir John had said. However, the fact is that the amount of radioactivity released at Windscale was about 4 orders of magnitude (10,000 times) less than that released in the atomic bomb at Hiroshima. These statements get into books, as this one has, get quoted, and the credentials of the statement are not there. I think it is very important that this subject of pollution should be dealt with on a sound quantitative basis.

As has been said by our last speaker, we may require in certain instances and for certain quantities to use the sea for certain purposes under controlled legalised conditions, and numbers matter in this case. It is terribly important not to quote generalised statements which could be at fault.

Dr Jean Carroz

I would like to make a few comments on the very interesting address given this morning by Dr Bowett. He told us that international agreement to regulate and prevent radioactive contamination of the sea was extremely limited so far. A few complementary remarks may be in order.

First, on the negative side, it may be mentioned that the 1963 Treaty Banning Nuclear Weapon Tests contains an important exception. The conduct of tests beneath the sea-bed is still authorised under Article 2 of the Treaty, provided, however, that no radioactive damage is caused to the environment, including superjacent waters.

On the positive side, it seems to me that several measures ought to be mentioned in addition to those listed by Dr Bowett. At the Geneva Conference of the Committee on Disarmament last October, a draft treaty was submitted by the USSR and the USA. Under that treaty, the placing of nuclear weapons on the sea-bed would be prohibited, which would give a further measure of control over the contamination of the sea. At the moment, negotiations are going on regarding supervision and enforcement methods.

Reference can also be made to the 1959 Antarctic Treaty, which contains a specific provision (Art. V-1) prohibiting any nuclear explosions in Antarctica and the disposal there of radioactive waste material.

As regards the control of disposal of radioactive waste into the sea generally, it is perhaps useful to describe briefly the work carried out by the IAEA. Reference was made this morning to the technical panel which met in 1960 and produced the Brynielsson Report. Following the meetings of that panel, the Agency convened a panel on the legal implications of the disposal of radioactive waste into the sea. The panel met several times between 1961 and 1963 and formulated two sets of draft Articles. The first one was rather straightforward, since it simply prohibited the disposal of any radioactive waste into the sea. It had been proposed by two countries which were absolutely opposed to the use of the sea as a dumping site. The second set of draft Articles was more elaborate and did allow the introduction of radioactive waste into the sea under certain conditions. These conditions are generally in line with the requirements that Dr Bowett suggested for any future action: obligation for states to report intended disposals to a competent international agency; maintenance of a register by that agency; designation of suitable sites; monitoring, etc. In view of the fundamental difference between the two sets of draft Articles, it has not proved possible so far to reach agreement on a single text, either allowing disposal under specific conditions or prohibiting any disposal. But I understand that negotiations are going on and that a consensus might emerge.

Finally, it may be relevant to mention that in accordance with resolution 2566 (XXIV) adopted last December by the General Assembly of the UN, the Secretary-General of the UN sent an enquiry to all member nations requesting them to advise him, by the end of next month, of their views on the desirability and feasibility of an international treaty or treaties to prevent and control marine pollution. As the resolution refers specifically to radioactive materials, it may very well be that, as a result of this enquiry, some new action might be taken.

MR J. E. S. FAWCETT, CHAIRMAN
I wonder, in the light of what Dr Carroz has said, whether we could ask first of all what national control of excessive disposal of radioactive waste exists in this country. Could we know what is the mechanism by which disposal that goes above the acceptable level is controlled?

MR A. PRESTON
Prosecution, under the Radioactive Substances Act, 1960.

MR J. E. S. FAWCETT, CHAIRMAN
How many prosecutions have taken place?

MR A. PRESTON

None so far as waste disposal is concerned; I think there have been no contra-
ventions of regulations. In so far as the misuse of sealed sources of radioactivity
are concerned, which are also controlled under the regulations, there have been
a number of prosecutions. There have also been a number of prosecutions in
relation to the working conditions in various parts of the luminising industry,
but as far as controlled disposal of radioactive waste is concerned there have been
no prosecutions to my knowledge. If there were, however, there is provision to
take the offenders to law under the Radioactive Substances Act of 1960.

MR J. E. S. FAWCETT, CHAIRMAN

What are the penalties?

MR A. PRESTON

I could not quote you the figures, but, clearly, fines of a considerable size may be
involved.

DR D. W. BOWETT

It may be easy in Britain at the moment because perhaps all the users of atomic
substances likely to produce waste are known, but if this usage becomes much
wider, the day might arrive when one finds that someone has disobeyed the law
and one does not know whom.

MR A. PRESTON

There is an absolute requirement for registration for the use of radioactive
materials of whatever kind throughout the United Kingdom.

DR D. W. BOWETT

Yes, but if one is monitoring a system whereby someone or somebody is putting
waste into an estuary or into the sea, and if one finds one Monday morning that
there is in the sea, or in the estuary, more radioactive waste than the limit
imposed allows, how does one know who is responsible?

MR A. PRESTON

The answer is, of course, that the number of installations putting radioactivity
into the sea or into rivers in quantities which could lead to this sort of situation
is relatively small – it is only the major users, the nuclear power stations, the
Atomic Energy Authority sites. The sites are well known, the points of discharge
are well known, and monitoring is extensive, so it would be easy to identify the
particular offender.

DR W. R. P. BOURNE

At the moment. But is there a technical means available in the future? If, for
example, there were atomic power stations in Southern Ireland, and then one
found that the concentration of waste in the Irish Sea was too high, which side
did it come from?

MR A. PRESTON

Well, we have a very good understanding of the general nature of the distribution
of radioactivity in the Irish Sea and we have very well established quantitative
relationships between the concentration and distance from the various outfalls

in the Irish Sea, so it is quite possible to trace back to the outfall which has given rise to the enhanced levels of radioactivity.

PROFESSOR J. F. GARNER
Could I follow up on that? What happens if there are two discharges into one estuary, and it is only because there are the two that you get above the minimum level, or whatever the proper expression is; whom do you prosecute? Both of them? The last one, or the first one?

MR A. PRESTON
Well the answer is, of course, that when you look at a river estuary like the Severn, for example, where there are already three operating power stations – a fourth under construction and a fifth under consideration – it is the capacity of the estuary as a whole that is assessed, and it is divided up in such a way as to not only cater for the present users but for the presently planned production of waste from nuclear power in that estuary. This would be the mode of operation in any other estuary.

PROFESSOR J. F. GARNER
But if, on monitoring, you found it was above the level . . .

MR A. PRESTON
But we shall not find that it is above the level, because we have authorised the rate of disposal which will keep it below the level, and it is only in contravention of that authorisation that it can go above the level.

PROFESSOR J. F. GARNER
But you see we are assuming that there has been a breach of the law.

MR A. PRESTON
But if there is a breach of the law, the offender concerned will be punished.

PROFESSOR J. F. GARNER
But how do you know who is the offender?

MR A. PRESTON
Because we can trace back by studying the pattern of concentration against distance from the outfall which is responsible for the discharge. Do not forget that the input of effluent is also monitored, and it is monitored not only by the operator but independently by the authorising departments who take their own samples.

DR W. R. P. BOURNE
Could I say something I feel rather strongly about in relation to all forms of pollution control? One cannot suddenly introduce regulations to control things without a lot of study first, and when one considers the control of various other major international abuses there needs to be some sort of preliminary method. The thing I came up against recently, in another context, was the original control of the slave trade, which was not suddenly brought about overnight. The first stage was registration of all the slaves – little else was done at first but details were obtained of what was going on. The second stage was control of the

further importation of slaves, and it was only after a number of years that the existing slave holdings were liquidated and at that point it was necessary to introduce compensation. I think that the big problem in pollution control is, first of all, to discover what is going on, which means that all sources of pollution will have to be registered, so that we have a list of precisely who is doing what, how much they are putting into the sea and who is the potential menace. The next stage, one would have thought, is to stop any further large scale investment which is going to cause pollution without built-in safeguards, and it is only after that that we can start to do anything about existing installations. It is liable to be very expensive indeed if we are suddenly going to dictate that nuclear power stations cannot release any more radioactive material, but the idea that we can suddenly introduce regulations overnight and control something about which we have no accurate statistics and which is going to entail expensive complications if we have not got a monitoring apparatus to discover whether it is having any effect or not, is just unrealistic.

MR A. PRESTON
This clearly does not apply to the radioactivity field.

DR J. W. HOPTON
So far we have overlooked what I consider to be an important point. When one is discussing legislation and standards one has to remember that this material is being passed eventually into natural biological systems, which can differ enormously in their properties. For example, if you consider a cold water lake in North Finland this is a very different biological system from the sea off the coast of Sicily, and there obviously would have to be different standards for different situations. How does one legislate for this kind of thing?

MR A. PRESTON
This was the very point which I hoped to have an opportunity to investigate, following Dr Marley, because it has a very pronounced bearing on what Dr Bowett was saying this morning about drawing up safe levels. Could I perhaps spend a few moments on it. You recall this morning the diagram of Dr Marley's on the acceptable level concept. Well, in terms of the amount of radioactivity you can tolerate in the body, the ICRP has established for a whole range of different types of radionuclide the quantity of concentration in an organ which will produce that acceptable dose, and it has also established the rate of intake, on a daily basis, to establish that burden and to maintain it but not exceed it. These are called maximum permissible daily intakes. These can then be interpreted in any individual situation in the light of the rate of consumption of, for example, laverbread or oysters or trout flesh or any other food material, and a level of intake can be set such that this organ burden can not be exceeded and, therefore, that the dose can not be exceeded.

The point I wish to make is that when you go to any individual environment you will find that the sequence of events which leads to the contamination of any given foodstuff is extraordinarily complex. You have differing degrees of dilution, you have differing degrees of removal on suspended matter, differing degrees of re-concentration in organisms, depending on the radionuclide, and depending on the organism. So one ends up specifying permissible concentrations in a whole range of materials in particular contexts. The concentration, for example, of zinc in oysters permitted in the Blackwater estuary would be quite different from the concentration of zinc permitted in oysters in another situation, because of the specialised features in the Blackwater environment.

It is therefore not possible, in scientific terms, and this is one reason why the legal technical panels convened after the first Brynielsson meeting did not really get very far. It is not possible, in scientific terms, to specify maximum permissible concentration in sea water which would be all-embracing.

What it is possible to do is to establish a code of practice along the lines of the ICRP and the critical path analysis to ensure that each situation will be treated on its merits and so as to ensure that acceptable levels will not be exceeded for particular situations. What we need is adherence to a code of practice. Now, IAEA is reconvening a panel of experts in November this year to look again at the Brynielsson Report and to see what new knowledge has come to light in the intervening decade to see how much further we can go towards a scientific basis for some sort of international control. I think, and this is a personal opinion, that although we have learnt a lot more about the behaviour of radioactivity in a marine context – or for that matter in a fresh water context – we are not going to be able to take the scientific considerations any further in terms of helping out a legal framework.

In terms of global monitoring, there are a number of international bodies considering it, and, as part of the Pre-Conference Seminar at the forthcoming FAO Symposium on Marine Pollution in Rome, there are several expert panels being set up to consider various aspects of marine pollution, including radioactivity and including the design and development of a global monitoring system. One prerequisite for global monitoring is, of course, a proper framework of internationally accepted methods of analysis of the materials you are trying to identify, and in this context IAEA has also set up in 1968 an expert group to recommend reference methods for the study of marine radioactivity, and to promote intercalibration excercises so that we may know with what confidence we may depend on the measurements of one country as against those of another. When we have an internationally acceptable set of methods on which to analyse the radioactivity, and an international Code of Practice as to how to introduce it into the sea, it will be a very large step in the direction that Dr Bowett would like to see.

Mr J. F. Whitfield
I am interested in Dr Preston's remark that nuclear power generation in the Severn Estuary will be extended and that a fourth station is being proposed. As a member of the Severn River Authority I have been involved in the proposals for this 600 megawatt nuclear power station at Stourport, which will be the only such station in the country, if it is finally agreed on, to be on non-tidal water – there are two, I think, in Germany. My Authority have put in a holding objection to the station because we have asked the CEGB to give us a guarantee that there will be no radiation pollution, but, as a layman, not as a scientist, I am

asking why the CEGB cannot give us a guarantee that there will be no radiation pollution in a non-tidal area like this.

Mr A. Preston

I am not a member of the CEGB and do not pretend to speak for them but, of course, it is very difficult to operate a nuclear power station without producing some radioactive waste material of either a gaseous or a solid or a liquid nature or, indeed, all three. I think what is important in the context of the Stourport proposal, is that if radioactivity of a liquid kind has to be introduced into the Severn that it can be introduced at levels which are acceptable. If they are acceptable, then, in practical terms, we can regard this as not polluting the Severn. But, of course, as you know, the Stourport proposal is at the moment at Government level for decision, and nobody knows at this stage which way the decision is likely to go, because there are more considerations to be taken into account than the question of waste disposal. The various factors that the Board has to take into account in siting a nuclear power station range from the cost of waste disposal to the siting in relation to existing coal-fields, transmission lines, the source at which the power is required, etc. – all these will have to be balanced up before a final judgment can be made, and that judgment is, of course, the remit of the Ministry of Technology on the advice of the departments concerned.

Mr J. F. Whitfield

I do not think I have had my question answered. The question is, why cannot the authorities give us a guarantee that there will be no radiation pollution possibilities in the river? Why cannot the potential radiation pollution hazard be sited elsewhere than in a non-tidal area?

Mr A. Preston

I think the answer to this question is that if it can be safely introduced into the river without infringing the acceptable standards it should be introduced into the river rather than incur an additional financial penalty in taking it elsewhere. There is no justification in doing this in terms of the acceptable standards of introduction. But, as I said before, I am not a member of CEGB.

Professor J. F. Garner

Is Mr Whitfield correct? Has not Harwell always discharged into the Thames fresh water?

Dr E. Windle Taylor

Could I shed a little light on the present discussion. A meeting is to be held among the scientists of water undertakings that are involved in the Stourport proposal, particularly on the question of any risk of radiation from low-level radioactive discharges but also from the point of view of thermal pollution, and on the wider issue of abstraction of water for cooling purposes. Limiting what I have to say to radioactivity, I think the way in which we may be able to obtain an answer to the question as to whether there will be a guarantee about the limits of radioactivity of the water is through the authorising authority, namely the Ministry of Housing and Local Government. I am Chairman of this particular Committee of the British Waterworks Association and I shall advise that way. In the past the authorisations have come from the Ministry of Housing and Local Government, and they will be able to give us more detail about how this nuclear power station is going to be planned. I know, and I agree with Mr Preston, that it is in a very early stage at the moment, but I can say that the water undertakings are very worried about it because it is the first inland nuclear power station contemplated in the country. Up to the present moment thay have been sited on the sea shore.

May I turn to another point with regard to Harwell? Harwell was the first large nuclear reactor station in this country and was installed in the late 1940s. Naturally, at that time, the water undertakings and the River Authority, the Thames Conservancy, were most concerned, but a very strict authorisation was applied to the discharge of the low-level radioactivity from Harwell, and here arises the point about independent sampling which was referred to this morning, I think by Professor Burhop. Three organisations sample every discharge of low activity liquid from the tanks at Harwell. There is the Harwell authority itself, there is the Laboratory of the Government Chemist, and there is my own laboratory, and I can say, with my hand on my heart, that since 1949 the authorisation has never been exceeded. I think you can take this experience as a successful example of control of radiation.

Dr J. A. G. Taylor (Unilever)
I wonder if we could get this subject into perspective and question Dr Marley in particular. We have been talking about changes in radioactivity, and the levels we are facing; I wonder if he could give us any idea of what the background was before the nuclear era, what it is now and how the graph is going.

Mr J. E. S. Fawcett, Chairman
If I could just intervene for a moment – perhaps this question could be answered in terms of control. It seems to me that the safe acceptable concept hides, in fact, certain social and political judgments and it may be that the Severn Authority does not find acceptable what physicists or biologists may find acceptable and for quite different reasons. "Acceptable" is a very variable term.

I think, if we are fastening on to the point of control, that what Mr Windle Taylor said points to the question whether the control should be up-stream or down-stream. What he seemed to be suggesting was that real control lies at the beginning of the operation, as it were, even in the licensing of the plant, rather than in the policeman coming along and saying that if you put too much radio-active effluent in, we shall punish you – that is what I call the down-stream control – and the IAEA seems to have done that to a certain extent inter-nationally; there is the notion that, if the USA sells nuclear plants to other countries, control is built into the contract. Is this something we could, perhaps, consider, particularly internationally. What kind of control are we talking about? Licensing control? Control of the plant? Or simply waiting to see what happens?

Dr D. W. Bowett
I cannot answer your question because I think you need a scientist to answer it, but it does concern me as a lawyer because the difficulty about moving your control to the land is that as you come in from the high seas and you move into

a state's territorial waters or, even more, into its national waters and then on to the land, so you are going to get increasing opposition by the states to any international control. They might be quite happy to have an international inspectorate on the high seas but they will be damned if they will have you walking around their nuclear power plants.

MR J. E. S. FAWCETT, CHAIRMAN
But doesn't IAEA have inspectors?

DR D. W. BOWETT
Not to my knowledge. In any event if they have inspectors they can only operate in the plants which are receiving materials through IAEA, and that is a very small minority. Perhaps someone can tell us: is there any international inspection in the UK power plants?

MR A. PRESTON
No.

DR D. M. MILLER
There is an area of international inspection but it is not in the pollution field, it is in the disarmament field and there will be more when the NPT becomes more operative, because the IAEA will be the agency which will be largely responsible for inspecting nuclear installations to make sure that none of the materials is being used for weapons.

MR J. E. S. FAWCETT, CHAIRMAN
But if you have this partial machinery already for weapons control, why cannot it be extended?

DR D. M. MILLER
The difficulty really is whether countries will be willing to recognise pollution as a danger on a par with nuclear weapons proliferation, and so will be willing to give up a certain amount of their sovereignty to subject themselves to that sort of control.

DR D. W. BOWETT
Do the scientists accept that national control is just not adequate. I do not know; they may think it is.

DR W. G. MARLEY
It is important to recognise that the release of radioactivity at a point on the coast is much more likely to produce trouble locally than it is to contaminate the high seas. For instance, we are slightly concerned in the UK about what the French are doing in releasing radioactivity at Cap de la Hague, but only mildly so. If they produce a considerable release there they are much more likely to affect their own inshore fishermen and their fish contamination level, so, from the point of view of expediency, their national control should safeguard the position with regard to international waters. The dilution as you move away from a local source is enormous; I think we can say that, roughly, a kind of inverse square law applies. I think that this has a bearing on the question of whether an international inspectorate is required for releases, point releases, from coastal sites.

The position will, of course, be different with regard to any disposal of radioactive wastes on the high seas.

Mr J. E. S. Fawcett, Chairman
Would that suggest, then, that even if there were an international authority, for example IAEA, control, in this sense, could be regional at the most and perhaps even national for most of the way?

Dr W. G. Marley
Could I just come back on one further point in regard to what Dr Windle Taylor said on discharge of radioactive waste. The comment was made that it was controlled by the user, that is, controlled at the operation of the plant. From the point of view of expediency this is undoubtedly the best way to do it. It is far more complicated if you look at waters down-stream and say, "That plant higher up-stream is discharging too much". This control, while it is scientifically interesting, is not so easy. It is much easier to effect a control on the amount put in at the point of discharge, and, therefore, coupling the control with the licensing is undoubtedly the right way to do it and this is a task for the national government.

Mr P. Armstrong
We have not been told how far out the nuclear waste is put when it is discharged into the sea.

Mr A. Preston
I think I did mention this morning in the case of Windscale that it reaches the sea through a mile and a half pipeline. The length of this pipe, of course, varies from site to site, depending on the particular hydraulic conditions, the nature of the environment and so on. It is a question, again, of judging situations on their individual merits and requirements. So far as UK practice is concerned, there are designated disposal areas for the disposal of packaged solid radioactive wastes in the deep sea. These disposal areas are selected on the basis of their remoteness because of the need to ensure that there can be no accidental recovery of the material dumped. This is achieved by dumping in depths of water in excess of 3,000 metres at distances of at least 50 nautical miles from the edge of the continental shelf in areas where weather permits reasonable conduct of the dumping operation and which are subject to acceptably accurate navigational coverage free from submarine cables. In fact, when one looks at the western approaches to the United Kingdom, and puts on the map the tracks of the various cables and the 1,000 fathom line, and then adds to the track of the cables an acceptable margin for navigational error, there are remarkably few areas where one can put solid packaged radioactive waste. There are three which are considered to be within acceptable economic steaming distances of the UK ports of lading and these are all some 7–800 miles south-west of the British Isles in the general area of the north-west European basin.

One of these areas has also been used for a joint European Nuclear Energy Agency dumping operation conducted in 1969 which is the second such operation – the previous one having been conducted in a disposal area off the Iberian Peninsula, in which many of the European nations joined in an operation along the lines I have indicated, and observing the same rules of procedure and so on. I might add that the amounts of radioactivity involved here are very small, much smaller than those delivered by pipeline into the Irish Sea. They are put into containers that are designed to reach the sea-bed intact and they have an average life on the sea-bed of about ten years. In terms of the consideration of the oceanographic factors which have to be taken into account the object is, first of all, to achieve disposal in a remote area and then to put in between this remote area and man, in terms of exploitation of marine resources in a contemporary context at least, an enormous dilution factor which is achieved as the waste material is released from the can and is returned to the surface waters. The sort of calculations that have been done show that the factor in hand between an acceptable level in waters presently exploited and the level that might be achieved from disposal operations conducted on the present scale is somewhere in excess of 10^8.

Lord Simon

Do I understand that when the container finally breaks up after 10 years the material comes to the surface? I would have thought it stayed at the bottom.

Mr A. Preston

It depends on the nature of the material. Clearly, some of this material is literally contaminated spanners and bits of equipment and things of this kind which it is not easy to treat in any other way; some of it is material which could go into solution through corrosion or directly – it may be in slurry-like form. That which does get into solution will clearly mix with the water and, since the deep water mass is returned to the ocean surface on a pretty long time-scale, probably with a half-life of about 100 years or in excess of that, you have a very long time for radioactive decay as well as actual physical dilution. One of the difficult problems, perhaps, with the deep sea disposal of solid radioactive waste is to ensure the exclusion of permanently floating materials such as polythene, and in UK waste this is excluded, we do not allow it to go to sea.

Mr F. Macdonald

Are we not forgetting one thing? We talk about discharging acceptable levels and dilution, but are we not forgetting the extreme stability of this material? Have a look at the DDT situation where, in the North Sea, we have concentrated this with acceptable levels over a very long period into a now unacceptable situation, as a result of countries making decisions in isolation. The Scandinavian countries have banned the use of DDT in their countries, but it is still coming in huge quantities down the Rhine and from other areas. Is not this the real problem, that a national decision in isolation is of no use where you have got this extreme stability of the material? BPCBs have been dumped in the North Sea for thirty years now.

Mr A. Preston

First of all, to use the parlance of pesticide pollution, although the radioactivity is not biodegradeable it does have a natural radioactive decay, as Professor Burhop mentioned, and some of the radionuclides only have half-lives which are measured in days, tens of days or hundreds of days. Some radionuclides do have a longer half-life. Those that are very short, of course, soon reach an equilibrium

47

with the rate of introduction of the material, provided the equilibrium which is reached is within the acceptable levels. I think that is a satisfactory situation. The more difficult situation, of course, is with the very long-life material and very little of this indeed is put into the sea – it goes to the permanent storage area which Professor Burhop mentioned this morning.

One other thing I would like to come back on while I am talking about this, since several people mentioned it this morning, is the question of the accumulation of radioactivity through food chains. We know that the non-biodegradeable organochlorine pesticides and things of that kind move up the food chain and eventually end up in some terminal link in the food web at fairly high concentration. In the case of radioactivity, with very few exceptions there is differentiation against the radioactive nuclide as you go up the food chain, and you do not get increasing concentration. There are a few radionuclides where this does not apply and those are caesium, zinc and phosphorus, but with most other radionuclides there is discrimination – you do not get the accumulation that you do in the case of pesticides. The figures which were quoted to you this morning by, I think, the opening speaker, on concentration factors in the food chain in the Columbia River leading to the prodigious concentration in wild bird eggs – aquatic birds – were actually shown to be in error. What happens when you introduce radioactive material into a water-course, whether it is the sea or a river, is that you get a changing pattern of concentration as you get away from the discharge point, and if the discharge is intermittent you even get variation in concentration at a given point. Therefore, you have got to be awfully careful how you interpret the concentration in a biological system from spot observations taken in the water mass itself when the system is integrating the average concentration over a large area or a long period of time. You can often divide an awfully high figure by an artificially low figure and get a very high concentration. This is the case with the duck eggs.

Dr J. A. G. Taylor (Unilever)
I should like to ask you the question I asked before. That is, whether you could give us some perspective on the levels you have been talking about. What was the level of radioactivity that the general public were facing before the coming of nuclear energy? What are the levels now and how is it going? Is it going up, levelling off or going down?

Miss M. M. Sibthorp
Could I ask a question about something Dr Preston referred to. He quoted and questioned the figures given for the Columbia River experiment, but the Columbia River experiment, I understand, was an experiment carried out by a group of scientists in the United States and, presumably, they must be responsible for their figures.

Mr A. Preston
Yes, but scientists make mistakes. This figure was admitted to be in error by the team of people working for the GEC Biology operation at Hanford. They have looked at the situation again and they do not find these high concentration factors in the Columbia River biota. The high concentration quoted was, of course, related to the radiophosphorus which has a half-life of 15 days, and so it is a

very transient phenomenon. Radiophosphorus is not a fission product, it does not arise from nuclear power, but it arises from the peculiar situation which existed twenty years ago at the Hanford plant environs, where they cooled the reactors by passing river water through them and activated, by neutron activation, the phosphorus in the water. It was a very peculiar situation and this also has to be taken into account; it is one of those elements which is biologically important.

MR J. E. S. FAWCETT, CHAIRMAN
Could you now answer Dr Taylor's question?

DR W. G. MARLEY
The level of natural background radiation in most countries of the world is about 0·1 rad per year dose rate of 100 millirad. In some countries, such as the United Kingdom, medical procedures account for an addition of 20 per cent – 20 millirad – and in the United States something like 40 millirad per year. Dose rates from fall-out of radioactivity, averaged on a population weighted basis in the Northern Hemisphere, have added a total dose which is about 100 millirad – this is in round figures – and this figure is the total dose over all time, from all the explosions of nuclear weapons. The dose rate in any one year, of course, is very small, because this is predominantly the dose from two long-level radionuclides – which are fission products – strontium 90 – which contributes a dose to bone tissue – and caesium 137 – which has also about a 30 year half-life and contributes a dose to genetic tissue. So you see that all the tests in all the nuclear series have contributed about what one man gets in one year of natural living; it has in effect pushed him forward one year. The contribution from the development of nuclear power on this scale is extremely low. In the United Kingdom occupational exposure – that is to workers – contributes (I am going to change the unit to man-rad, that is the dose of one rad to one man), 10,000 units. This can be compared with the natural background which is 50 million times 0·1 rad (or 100 millirads), which is 5 million man-rad. So the occupation of workers in the Atomic Energy Authority causes a 10,000 man-rad dose, natural background causes a 5 million man-rad dose per year at the present time. So the occupational exposure is a very small contribution in relation to the natural dose of radiation incurred by the population. It is thought that other radiation workers, such as those in hospitals, contribute about as much again; so the development of atomic energy and radioactivity and radiation work at the moment contribute a negligible amount compared with the natural background dose.

There is one other thing I would like to say about this natural background. It varies from place to place and it is important, if you are considering reduction of the dose to the people in this country, to realise that in some towns the radiation background is double the normal – Aberdeen and St Ives in Cornwall rate for consideration here – and if you are considering the expenditure of money to reduce the background you must, logically, consider whether it would pay to, shall we say, rebuild some of these towns with non-granitic material. These are logically things you must consider if you wish to reduce the dose. On the other hand, as time proceeds, we are getting better in our understanding and evaluation of the effects of these very small doses, and I think we should, perhaps, hold our hand on these rather dramatic developments.

The David Davies
Memorial Institute of
International Studies
The Department of International
Politics of the University College
of Wales, Aberystwyth

CONFERENCE ON

LAW, SCIENCE AND POLITICS:
WATER POLLUTION AND ITS EFFECTS
CONSIDERED AS A WORLD PROBLEM

Saturday 11 July

AFTERNOON SESSION

Oil Pollution

Chairman	PROFESSOR P. F. WAREING University College of Wales
Scientist	PROFESSOR R. B. CLARK University of Newcastle upon Tyne
Administrator	MR MAURICE HOLDSWORTH Shell Petroleum
Lawyer	PROFESSOR R. Y. JENNINGS, Q.C. Jesus College, Cambridge
	(Paper read by DR WHITE University of Manchester)

The Biological Consequences of Oil Pollution of the Sea

PROFESSOR R. B. CLARK (Department of Zoology and Dove Marine Laboratory, University of Newcastle upon Tyne)

Coastal oil pollution has been a matter of public concern – or at least complaint – for half a century. Unlike most other pollutants, oil is visible even to the uninstructed eye and attracts immediate and constant attention. Its nuisance value is obvious and were fouling of tourist beaches the only consequence of oil pollution, policy with regard to control and treatment could be easily settled. However, oil pollution has other consequences, all of which are, in one way or another, biological. This introduces new problems and often a conflict between amenity and biological interests. Deciding upon appropriate measures to counter oil pollution then demands the assessment of priorities which will vary from place to place and even from time to time. Ultimately the matter is likely to be resolved on some kind of economic basis, but whereas it is not very difficult to evaluate the damage to a fishery or the cost to a tourist resort resulting from oil pollution and decide priorities on these grounds, it is much more difficult to place a cash value on scientific or aesthetic damage to the coastal environment and even more so to consider long-term and insidious damage in these terms. The last is the most serious. Not only is it almost impossible to assess prolonged damage in realistic economic terms, and for this reason it is likely to be ignored altogether, but long-term consequences of oil, or any other kind of pollution, cannot easily be detected until a stage is reached from which recovery will be extremely slow.

Biological problems are almost invariably very complicated because of the extensive interactions between organisms in an environment like the sea and it follows that none of the answers will be simple and most will be hedged around with qualifications. Furthermore, because of the relative neglect of the ecological sciences in the past, there are considerable areas of ignorance so that any discussion of the impact of oil pollution on biological systems is inevitably a mixture of fact and supposition. There is a considerable and understandable temptation to the layman to dismiss biological considerations because of their complexity and apparent inconclusiveness. It is wrong to succumb to this temptation, however; the very fact that we are ignorant of the consequences of our activities should be a matter of great concern. We are becoming more conscious than ever before of our dependence upon the natural environment of which we are a part, but at a time when we are forced to exploit it as fully as possible we also have the capacity to inflict damage upon it on such a scale as to have serious repercussions on our own welfare and livelihood. This is a situation in which we must act as good husbandmen to the whole natural environment and to do this we

53

must not only seek to understand the consequences of what we do, but also to take warning from the earliest signs of damage and find out whether or not it is supportable.

Seabirds

Almost the only direct and significant casualties of oil pollution at sea are diving seabirds, more particularly auks (guillemots, razorbills and puffins), penguins and some diving ducks. These are birds which spend the greater part of their lives on the water, swim rather than fly (penguins are flightless and auks are poor fliers) and hunt their food underwater. If, as commonly happens, they surface through an oil slick, they become completely covered with oil in a way that other, flying birds are not. Furthermore, these birds are highly gregarious so that if they encounter an oil slick many individuals are likely to become contaminated. Although the incidence and extent of oiling is most severe in these diving birds, all birds that settle on the surface of the sea and those (principally waders) that feed on estuarine mud flats and the shore are liable to suffer contamination of the plumage by oil when they encounter an oil slick or a polluted beach.

The matting of the plumage by oil allows water to penetrate the air spaces between the feathers and the skin, with the result that the birds lose buoyancy and also, having now no natural insulating layer of air, they rapidly lose heat. To counter this, there is an increased metabolism of food reserves in the body amounting to twice the normal rate in heavily oiled birds and showing a substantial increase even in birds with only a small patch of oil on their feathers (Hartung, 1967). This increased metabolism is not balanced by a corresponding increase in the dietary intake. Indeed, a severely oiled bird is unable to hunt and catch food and, as a result, rapidly becomes emaciated as well as cold and ill, so that its chances of survival are small.

The causes of death are probably multiple. Many severely oiled birds may become waterlogged and drown (Tåning, 1952); the extent of mortality from this cause will obviously depend upon the natural buoyancy of the bird, weather conditions and the distance from land. Commonly death may be due to emaciation: "wrecks" of seabirds in winter when prolonged stormy conditions prevent feeding at sea are not unknown and under these circumstances the birds are severely emaciated with considerable wasting of the flight muscles (McCartan, 1958), and a similar condition has been observed in many dead oiled birds (Beer, 1968a).

Damage caused by swallowing oil, which is inescapable when the birds attempt to preen themselves clean, is hard to assess. Most oiled birds received at cleaning centres are suffering from enteritis and this is responsible for the very heavy mortality during the first few days in captivity (Beer, 1968b). On the other hand, birds that have been allowed to preen themselves clean of oil have sometimes shown a remarkable recovery (Bourne, 1968; Hughes, in Clark & Kennedy, 1970). Complicating factors are the age of the oil (weathered oil from which the more toxic lighter fractions have evaporated is likely to be less damaging than fresh oil if swallowed) and whether or not it is mixed with toxic oil spill removers.

If crude oils encountered at sea are relatively non-toxic, the same is not true of refined products. Experiments carried out by Hartung (1963) and Hartung & Hunt (1966) on a variety of ducks have shown that a number of lubricating oils, cutting oils, diesel and fuel oils cause severe intestinal irritation and haemorrhage, diarrhoea, kidney and pancreatic damage, fatty degeneration of the liver, loss of mobility, loss of balance and loss of muscular co-ordination. Similar pathological conditions have been observed in autopsies of birds killed in oil pollution incidents (Hartung & Hunt, 1966; Beer, 1968a).

The extent of the mortality at sea from these causes cannot be assessed, but it is generally assumed to be high and that only a small proportion of oiled birds reach the shore. The numbers of oiled birds recorded in shore counts may represent as little as 10 per cent of the total number affected (Clark, 1969) and there is a possibility that this figure may include a relatively high proportion of birds that are heavily oiled (Bourne, 1968). This poses a serious problem for centres involved in cleaning the birds.

Cleaning and rehabilitation of oiled birds has been attempted in many places. Some species such as gulls, geese and swans, which do not generally suffer heavy oiling, can be rehabilitated with a fair measure of success. Normally they spend as much time on land as on the water and they respond well to captivity. This is not the case with the more aquatic, diving species. All methods of removing oil from the plumage at present in use destroy the water-repellent properties of the feathers. Although the birds may be cleaned and restored to health, they cannot be returned to sea immediately because the plumage rapidly becomes sodden and waterlogged when exposed to water, and the animals lose buoyancy and become cold. It is a long time, often as much as a year, before the plumage regains its natural waterproof properties and the birds can be released. To have to maintain birds in captivity for so long is expensive, makes great demands on the physical resources of the cleaning centres, and militates against success. Marine diving birds, in particular, are not equipped to spend long periods on land. They develop arthritis and are a prey to such diseases of captivity as aspergillosis and enteritis which are fatal to a large proportion of the birds.

It is becoming clear that successful rehabilitation of oiled seabirds demands a high degree of individual attention to the recuperating animal. Those people who treat a handful of birds at a time often have better success than centres which handle many birds, and the success rate of the larger centres is least when they are faced with a sudden influx of birds for treatment after a substantial oil spill (Beer, 1970). However, individual attention to birds, while it may assist recovery, introduces a new problem. A number of species of bird, particularly auks, appear to become tame quite readily and there is some evidence that it is difficult to re-introduce a tame bird successfully into the wild population (Tate, 1935; Bourne, 1968).

The impact of oil pollution on seabirds can be gauged more reliably by observing changes in the natural populations than by considering the number of birds seen to be oiled. On these purely biological grounds, the deaths of several species of gulls from oil pollution on British coasts is totally without significance because their numbers are increasing steadily despite the mortality from this source. On the other hand, oil pollution is certainly a contributory factor to the marked reduction in the numbers of long-tailed ducks (*Clangula hyemalis*) and velvet scoters (*Melanitta fusca*) migrating through the Baltic States (Bergmann, 1961; Lemmetyinen, 1966), to the decline in breeding colonies of puffins and guillemots in south-west England and Brittany (Milon, 1966; Parsloe, 1967a, b; Bourne, 1968) and of guillemots and razorbills in Newfoundland (Giles & Livingston, 1960; Hawkes, 1961). These changes have come about after some decades of losses by oil pollution and with the increased traffic of oil tankers around southern Africa there is fear that the jackass penguin (*Spheniscus demersus*) on the South African coast may suffer the same fate.

The greater impact of oil pollution on one species than another is not simply due to greater exposure or vulnerability, although that is an important factor. The consequences of oil pollution for the local or global population of a species can be assessed only by taking into account such demographic considerations as the age structure of the population, birth rate, death rate, age at maturation, and longevity. Auks, for example, have few natural enemies as adults and are long-lived; the world population can therefore be sustained by an extremely low birth rate. An increase in adult mortality such as that caused by oil pollution has a much more dramatic effect on the population of a species with these demographic characteristics than upon another characterised by high adult mortality coupled with a high birth rate. In the latter case increased mortality, particularly if it is intermittent, has little influence on the total population. An added complication in social birds, as many seabirds are, is that smaller colonies probably have proportionately less breeding success than large colonies (Darling, 1938). Hence if the number of adults in a colony is reduced by a local oil pollution incident, recovery, already slow because of the low birth rate, is further hindered simply because the colony is reduced in size.

Reduction of breeding success is clearly a more potent factor in determining population size than adult mortality and there is evidence that, besides the indirect effect resulting from a reduction in the size of colonies, oil pollution may have a more direct impact on reproduction. Spraying eggs with oil has been used for some time to control numbers of gulls and cormorants (Gross, 1950) and even slight contamination of eggs with oil reduces their hatchability in a very marked way (Hartung, 1965). An oil slick may therefore cause more deaths than is apparent from the mortality of adult birds. Rittinghaus (1956) reported a case of Sandwich terns (*Sterna sandvicensis*) which were lightly contaminated with oil that had washed ashore: eggs which were subsequently smeared with oil from the plumage of the brooding birds failed to hatch even though they were incubated for an abnormally long time. Breeding success of gulls contaminated with oil during the *Torrey Canyon* incident was also much reduced (O'Connor, 1967). There is also evidence that birds that have swallowed oil stop laying (Hartung, 1965).

Bearing these facts in mind, it is possible to formulate a sensible conservation policy for seabirds endangered by oil pollution. First, attempts to clean and rehabilitate oiled seabirds and return them to the breeding population are worthwhile only for species such as auks, penguins and some arctic duck, which have a low birth rate. Cleaning and rehabilitation of birds with a relatively high birth rate, such as gulls, will have no discernable influence on the population which,

56

in any case, is generally increasing. Second, it is necessary to consider only the recent human population explosion to realise that factors affecting the birth and pre-adult mortality rates have far more impact than anything else on the population. Conservation measures designed to increase the breeding success of species with a declining population would more than offset any losses from oil pollution. This should not be difficult in species such as the guillemot (*Uria aalge*) which, despite the low reproductive rate, suffers about 60 per cent loss of eggs and fledglings (Uspenskii, 1956; Tuck, 1960; Southern, Carrick & Potter, 1965).

Commercial Fisheries

The most productive areas of the sea are on the continental shelf and, with the exception of tuna fishing, these areas support the only fisheries of economic consequence. They are also the site of offshore oil fields, the areas with the greatest concentration of shipping and the severest navigation hazards, and they are the most subject to oil pollution.

Assessment of the actual or potential threat to commercial fisheries by oil pollution is difficult and beset with complicating factors. The hazard varies widely with the provenance of the crude oil or the nature of the refined product, with the biology of the commercial species at risk, and finally, much depends upon the hydrography of the area where an oil spill occurs, the weather and sea conditions prevailing at the time of the spill, and the measures taken to deal with it. Generalisation about the dangers of oil pollution to fisheries is almost impossible, and regulations and control of fisheries and anti-pollution measures are made more difficult by the intensely international nature of the problem.

The complexity of the problem can be seen in the Baltic, Mediterranean, and North Sea. These shallow, relatively enclosed bodies of water receive the effluents from all the countries bordering them, are as heavily trafficked by shipping as any seas in the world, and they provide important fishing grounds for the whole of Europe. The North Sea, for example, is fished not only by the countries bordering it but also by fishing fleets of the Soviet Union, the German Democratic Republic, and Poland. While finfish represent the most important crop, there are also substantial fixed inshore fisheries for bivalves (oysters, mussels and cockles), crabs, lobsters and prawns. Oysters and mussels, in particular, are often cultivated, so that a sizeable permanent industry is at stake.

The Baltic and North Sea are probably exploited to something approaching their limit so that any damage to the natural resources of these areas would have economic repercussions. Furthermore, both must be regarded to a considerable degree as single biological units so that damage caused in one area is likely to concern all countries bordering it. (Neither is true to anything like the same extent of the Mediterranean.) Thus the chief breeding grounds of plaice in the North Sea are in the shallows off the Dutch coast, but the most important fishing grounds are in the international waters on the Dogger Bank. Destruction of young fish on the Dutch coast would therefore have a widespread and considerably delayed impact.

Crude oil and, to a lesser extent, refined petroleum products are a complicated mixture of hydrocarbons, mainly paraffins, naphthenes and aromatics, and their derivatives. They encompass a considerable range of molecular weights, boiling points and solubilities in water. On the whole, it is the low molecular weight, low boiling point compounds that are the most dangerous. When fresh crude oil is spilled on the sea its composition immediately begins to change; volatile lighter fractions are lost by evaporation and others are lost by solution into the water so that eventually only the high molecular weight, non-volatile, insoluble fractions remain. The rate at which these processes take place depends upon sea temperature and weather conditions, but as a rule, freshly spilled oil is in varying degrees toxic, but after it has weathered a day or so its toxicity is much reduced. The rate of loss of toxic constituents is also affected by the behaviour of the oil on the water. Libyan crude oil, for example, thickens when poured on to water and does not form a thin even film as most other crudes do; this may be expected to reduce the rate of diffusion of water soluble compounds from it.

Refined products are simply distillates of crude oil, sometimes with minor additives. They therefore contain all the constituents of the crude over a given boiling point range. Light refined products consist almost entirely of the more toxic elements of a crude, and heavy refined products such as marine diesel oil, lubricating oil and heavy fuel oils, and residual oils such as Bunker C, contain very little of the water soluble toxic constituents.

These complications, coupled with the fact that crude oils from different oilfields vary widely in their composition, make it extremely difficult to interpret realistically the results of the toxicity tests that have been carried out in the laboratory. Interpretation is not made easier by the present lack of uniformity of test methods and test organisms used in the various laboratories engaged in this work. Nevertheless, certain general conclusions can be drawn.

As might be expected, fish eggs, developmental stages and young fish are more sensitive than adults to these toxins and it appears that, other things being equal, smaller organisms are more sensitive than larger ones (Smith, 1968; Nelson-Smith, 1970). Petroleum products at sub-lethal concentrations may also result in developmental abnormalities in young stages which ultimately lead to death in a number of animals. The presence of crude oil and other noxious substances in the water reduces the filtration rate of filter-feeding bivalves (Chipman & Galtsoft, 1949) and, if prolonged, this effect would be reflected in a reduction in their growth rate. To an unknown extent, these substances may also depress the growth of phytoplankton on which all animal life in the sea ultimately depends, either directly, or indirectly by reducing light penetration and depressing the rate of photosynthesis. There is also some evidence that exposure to sub-lethal doses of toxic petroleum products may prevent some animals breeding again.

It may be misleading to accept the results of toxicity tests simply at their face value. It is necessary to translate these laboratory findings into the vastly more complicated situation that prevails in the natural environment. There is, for example, a strong likelihood that many fish leave an area contaminated by oil or dispersants. It is known that petroleum products are repellent to fish (Blumer, 1969) and in the Santa Barbara channel at the height of the oil spill in 1969 there was no heavy mortality of fish although fish shoals were observed from the air by professional fish spotters in areas not contaminated by oil. Some of the older findings, as for example that fish exposed to oil are suffocated because the oil coats their gills (Wiebe, 1935), are thus probably irrelevant to the situation that actually prevails in the sea.

The most damaging oil spills are likely to be those close inshore and in shallow water. Shallow inshore waters provide important nursery and spawning grounds for many species of fish and are particularly vulnerable to the damaging effects of pollution of all kinds. This is not due simply to the sensitivity of young fish and developmental stages to toxic substances but to the reduced dilution of toxins and to the coherence of water masses.

The distribution and dilution of toxic additives to the water depends chiefly on mixing of water masses rather than to diffusion, but in many areas the rate of mixing is slow and discrete bodies of water may retain their identity for some time. There is also a separation of water masses on a larger, more permanent scale. In the North Sea, for example, there are several distinguishable bodies of water each characterised by different planktonic indicator species (Laevastu, 1960).

The consequences of these hydrographic conditions for marine pollution was illustrated in an incident on the Dutch coast after illegal dumping of copper sulphate near Nordwijk in 1965. The polluted water remained in a narrow coastal strip during the following two weeks as it spread 80 km. northwards towards the Waddensee, and its concentration at the end of that time was still 20 per cent of what it had been at the beginning (Roskam, 1965, 1966).

A floating oil slick does not observe hydrographic boundaries but travels with the wind at about $3\frac{1}{2}$ per cent of the wind speed and largely independent of water currents (Smith, 1968). The exposure of the water column to the oil floating on it therefore depends upon the relationship between wind speed and direction, and current velocity. Once an oil slick is emulsified or sunk, it becomes a part of the body of water and no longer moves in the same way.

Laboratory toxicity studies and analyses of the importance of local hydrographic conditions permit some assessment of the potential hazard of oil pollution to commercial fisheries, but whether or not oil pollution materially damages fisheries in practice can be decided only by considering fishery statistics in relation to the incidence of oil pollution. The North Sea and Baltic have been subject to continuing oil pollution for decades and it has been a matter of comment and concern since about 1920. In the half-century since 1920, oil pollution has been individually on a minor scale but in aggregate has probably been substantial. Nevertheless, finfish landings from the North Sea were three times greater in 1967 than in the early 1920s and 6–7 times higher in the Baltic (Table 1). There was some concern in the 1930s that fish catches were being reduced by overfishing but at no time has there been realistic concern or convincing evidence that oil pollution had any material direct impact on fisheries.

The largest and most damaging single oil pollution incident in European waters was the wreck of the *Torrey Canyon* in March 1967. In this case there was certain evidence of a loss of fish eggs and young stages from the area. There was a very high mortality of pilchard eggs in some areas affected by oil and young fish were scarce or absent in areas where toxic oil removers were used

Table 1

FISH LANDINGS (IN MILLION KG) FROM THE NORTH SEA AND BALTIC, 1920–67

	North Sea	Baltic
1920	1,194	78
1925	1,016	100
1930	1,380	93
1935	1,313	129
1938	1,429	140
1946	1,230	81
1950	1,530	125
1955	2,018	371
1960	1,553	406
1965	2,597	421
1967	3,030	498

(From *Bull. statist. Pêch. marit.*, Copenhagen)

at sea (Smith, 1968). Nevertheless, despite the heavy pollution in south-west England and the widespread, sometimes excessive, use of toxic emulsifiers in the area, there was no sign of damage to fish stocks and catches of whitefish during the ensuing season and subsequently have been at least the normal level.

It is now the widespread view among the European fisheries authorities that, with one or two reservations, neither oil nor oil spill removers present a hazard to commercial fisheries (Korringa, 1968; Simpson, 1968; Cole, 1969).

If oil pollution in the North Sea and Baltic appears to have been without appreciable consequence to fisheries, this has not been the invariable experience elsewhere. Most of the oil transported in European waters is crude oil, but the greater part of that in coastal waters of the United States is in the form of refined products which are generally more toxic and damaging. The fact that considerable stretches off the east coast of the United States are protected by offshore bars so that the seaways are essentially estuarine and enclosed, compounds the problem and oil spills in these waters have a greater impact on fish populations than in European waters. In September 1969, the grounding of the barge *Florida* resulted in the escape of about 2,000 barrels of No. 2 fuel oil at West Falmouth, Massachusetts, with an almost total loss of fish, molluscs, crustaceans and worms in the area (Hampson & Sanders, 1969; Blumer *et al.*, 1970). Oyster and soft shell clam beds that had taken three years to develop at Wild Harbor River were destroyed and large numbers of lobsters were killed. All shell fishing in the area had to be banned for an indefinite period. This extensive damage is unusual in Europe but it is not uncommon in United States waters; recovery from such an incident, particularly of relatively static animals such as oysters, clams and lobsters, is likely to take a number of years, although there might be an immigration of finfish to replace the loss of these fairly quickly.

Chronic pollution prevents recovery, however, and the impact of this on fisheries can be seen in the Caspian. The western part of the middle and southern

Caspian Sea is subject to intense and continuous oil pollution from several sources (Kasymov, 1970). Offshore drilling coupled with numerous natural and artificial oil seepages from the sea-bed account for 500,000 kg of crude oil per day escaping into the sea in the region of Neftyanye Kamni Island alone. Waste waters from refineries and the petrochemical industry contribute a further substantial amount of oil to the sea while the Caspian tanker fleet and fishing vessels are estimated to discharge some 900,000 cubic metres of oily water into the sea daily. In addition, large quantities of untreated sewage are discharged into the sea, but the chief pollutant is oil. The consequences have been dramatic. The production of phytoplankton on which all life in the sea ultimately depends has fallen markedly between 1962 and 1969 and this has been reflected also in a decline in zooplankton production. Since 1930, catches of sturgeon in the Caspian as a whole have fallen by two-thirds and in the southern Caspian are now only a quarter of what they were. Salmon catches have been reduced to a tenth their former figure. Catches of bream, carp, pike and sild have declined almost as much. The annual cost of pollution to the fishing industry in Azerbaijan is estimated at 2 million roubles and the losses of sturgeon in the Caspian are valued by the Central Scientific Research Council of the fishing industry at 35 million roubles per annum. It is true that lowering of the sea level in the Baltic and hydroelectric installations on the Volga and Kura have had some influence on fisheries but there is no doubt that if intense pollution continues on its present scale, the Caspian will become almost worthless as a fishery resource.

It is possible that as an enclosed sea, the Caspian presents a special case unlikely to be paralleled in more open seas, but it cannot be doubted that the exploitation of the North Sea oilfield exposes it to new dangers. Anxiety on this score has already been expressed by the Chairman of the Fisheries Committee of the Norwegian Parliament (*The Times*, 19 June 1970). Like the Caspian, the North Sea receives industrial and domestic effluent from a large population living around its shores; if to this is added a sharp increase in oil pollution its value as a fishing ground, too, may be damaged.

Although crude oil, particularly floating as a slick, has relatively little impact on fisheries except when pollution is severe and continuous, oil pollution and the measures used to deal with it can cause considerable indirect damage to commercial fisheries by tainting. Phenols and related compounds impart an unpleasant flavour to fish, which is detectable by humans even at extremely low levels of contamination, and tainted fish are unmarketable (Martin, 1970). Crude oil is not the only source of contamination; almost any petroleum product may cause tainting and in studies conducted by Surber in Ohio, it was found that outboard motor exhaust derived from burning 2·6 (US) gallons of petrol–oil mixture was sufficient to taint fish in one acre-ft of water (Tarzwell, 1970). Many of the older oil spill removers are applied in a kerosene solvent and this taints fish and shellfish at least as effectively as the crude oil it is designed to counter.

The fishing industry is vulnerable to the vagaries of mass psychology in a way that few others are and even the suspicion that fish may be contaminated is sufficient to depress the market.

During the period when the *Torrey Canyon* was in the news, fish sales on the Paris market dropped by half, irrespective of the quality or origin of the fish

(Korringa, 1968); during the outbreak of paralytic shellfish poisoning in Newcastle upon Tyne in the summer of 1968, sales of whitefish fell dramatically despite the fact that stringent controls by the public health authority prevented affected shellfish reaching the market and at no time during the outbreak were detectable amounts of toxin found in any commercial food fishes (Clark, 1968; Wood, 1968).

Adult finfish are unlikely to be contaminated directly by crude oil for if conditions become too unpleasant the fish move to other areas. The chief risk comes from fouling of fishing gear, a particular danger if an oil slick is treated by sinking, and this may lead to tainting of the whole catch.

Shellfish may be affected in a variety of ways (Simpson, 1968). Stranded oil contaminates the shells of molluscs without damaging them, but even a small quantity of oil on a few shells of mussels, cockles or winkles can taint the whole batch if, as is commonly the case, they are cooked before being shelled. Winkles (*Littorina*) from Poole Harbour were unsaleable for several months following an oil spill there for this reason. If oil is partly emulsified by wave action or oil spill remover, the fine globules may be trapped and ingested by filter-feeding animals which are then tainted and unmarketable even though otherwise unharmed. This happened once in Morecambe Bay with the result that collecting mussels (*Mytilus*) had to be suspended for two weeks until the animals had cleaned themselves. Oil-contaminated oysters (*Ostrea*) in the United States took two months to lose their oily taste (Mackin & Sparks, 1962).

Persistent low-level contamination can also cause tainting of shellfish. Clams (*Mercenaria*) are harvested as a commercial crop in the upper part of the Southampton Water which is subject to mild but chronic oil pollution, but the clams have to be relaid in unpolluted water for a time before they can be sold.

In summary, the ways in which commercial fisheries are damaged or hazarded by oil pollution are as follows:

(a) Damage to fixed fisheries by killing the commercial species or, more commonly, by contaminating them so that they cannot be marketed for a period. At present it is principally shell fisheries that are at risk but conceivably in the future fish-farming installations would be exposed to the same hazard. Since the fisheries are in fixed sites it would generally be technically possible to provide permanent protection for them, but this is likely to be prohibitively expensive.

(b) Damage to fisheries by spills of highly toxic refined products in relatively enclosed bodies of water. At present, because of geography and the different patterns of coastal tanker trade, this appears to be a particularly acute problem on the eastern seaboard of the United States, but not in European waters.

(c) Damage to particularly vulnerable areas such as breeding and nursery grounds of commercial fish. The consequences of widespread damage in such an area would be detectable for some years.

(d) Persistent and heavy pollution around offshore oilfields and industrial installations where, if pollution control is not strict, the entire productivity of the marine environment may be reduced to the point that fish, as higher members of the food chains, are seriously reduced in numbers.

Coastal Flora and Fauna

Detailed surveys of the damage caused to coastal plants and animals and their subsequent recovery have been made after several major oil spills. These were in a small cove in Lower California where the *Tampico Maru* was wrecked (North *et al.*, 1964), in south-west England after *Torrey Canyon* (Nelson-Smith, 1968; Smith, 1968), at Santa Barbara (Straughan, 1970; Straughan & Abbott, 1970) and in Buzzard's Bay, Massachusetts (Blumer *et al.*, 1970). The last two accidents were too recent for much information about recovery from the damage to have been gained yet. Less extensive surveys have also been carried out after a number of other oil spills. All these investigations have confirmed one another generally and we now have a reasonably clear impression of what consequences to expect from oil spills in shallow water and from stranded oil.

In south-west England and at Santa Barbara, the pollutant was crude oil which proved to be not particularly toxic in itself. Indeed, in the *Torrey Canyon* incident it is likely that more damage to animals and plants was caused by the misuse of toxic oil spill removers than by the oil itself. A certain number of animals were killed by smothering with thick layers of oil and "chocolate mousse" and some seaweeds with dry fronds to which the oil became attached were so heavy that they were torn from the rocks by wave action, but long-lasting disturbance of the ecological balance followed the destruction of herbivorous animals on the beach, principally gastropod molluscs. This was also the experience in Lower California following the wreck of the *Tampico Maru*. In this case the pollutant was not crude oil but far more toxic diesel fuel oil and this resulted in the death not only of herbivorous gastropods but also the death or disappearance of sub-tidal sea urchins which are also herbivores.

The biological character of a rocky beach depends upon the balance between the settlement and growth of seaweeds and the activities of the herbivores. The weeds produce spores which are widely distributed and settle to product sporelings on any solid surface. These young plants are grazed by winkles, limpets, abalones and the like and, if the intensity of grazing is sufficient, the beach is kept relatively free of weed and is populated instead by animals such as barnacles and mussels which are attached to the rocks in their place. If grazing is not very intense the weeds become established and exclude the mussels, barnacles and limpets to a considerable degree.

This is what happened in the bay in Lower California after the oil spill from the *Tampico Maru* and on some beaches in Cornwall where oil spill removers had been used extravagantly. In the former case, the bay became choked with giant kelp (*Macrocystis*) and it was some ten years before a gradual return to the former condition began. On the Cornish beaches that were affected, the rocks became covered with slippery green weed (*Enteromorpha*), later succeeded by brown wrack (*Fucus*), and it is only now that a few limpets are beginning to establish themselves on the beach and starting to graze down the growth of weed.

These are the most obvious changes but not the only ones. A weed-covered beach harbours a very different fauna from bare rock and the effect of the pollution is to change the entire ecological character of the affected areas. As far as we can tell, recovery of the original situation will be almost complete after a number of years – perhaps a decade or more. Some rare species may not reappear (Nelson-Smith, 1970) but although this is of some scientific concern, it is not ecologically significant. The evidence we have suggests, as might be expected, that the more toxic the pollutant, the more severe the kill and the slower the recovery.

Chronic pollution is a different matter. Even the plants are not allowed to recover from their initial setback and the polluted area becomes more and more impoverished of plant and animal life until only a few of the most resistant forms remain. The effect of repeated spraying with crude oil on salt marsh plant communities has been studied in detail by Baker (1970) at the Oil Pollution Research Unit, Orielton, Pembrokeshire. While a single spraying with oil may cut back some plants, recovery in the following season is generally good and growth may actually be enhanced. Damage to annual plants before they have seeded is, of course, not followed by such rapid recovery. Repeated spraying, however, eliminates almost all plants; only the Umbelliferae survive and they are known to have a great tolerance for oil.

Treatment of Oil Pollution

Avoidable oil pollution may eventually be completely prohibited and, depending upon the effectiveness of persuasion, policing the seas, and punitive action against offenders, may be eliminated, but accidental oil spills will continue so long as oil and oil products are transported by sea, even though we may hope that they will become less frequent. Whatever national and international legislation is introduced, therefore, we must expect continuing oil pollution and must have effective means of dispersing, sinking or removing oil slicks at sea and treating polluted beaches.

It is now general experience that it is simpler to deal with spills while the oil is floating at sea than after it has come ashore, though because it is not always possible to mobilise sufficient ships or aircraft to deal with floating oil, effective means of cleaning beaches are still needed. Two methods of dealing with floating oil slicks appear to be promising: by dispersing them with emulsifiers (oil spill removers) or by sinking the oil, although both methods have their disadvantages. Other methods such as the mechanical containment of the oil with inflatable booms and the like, recovering or scavenging the oil by physical methods, and consolidating it or converting it to a gel as an aid to scavenging, are all in use. These suffer from the disadvantage of being useful only in special circumstances or of being very expensive. Beach cleaning is carried out by spraying with oil spill removers or by the mechanical removal of oil, but where possible it now seems preferable to leave the beaches untreated and to allow the oil to degrade naturally.

Cheap and effective emulsifiers commonly employed as oil spill removers are used in a kerosene or kerosene extract solvent; as such they are more toxic than the oil they are used to treat as well as being unpleasant for the operators to handle. New, water soluble and much less toxic emulsifiers have now been developed and are gradually coming into use.

64

Although treatment of oil slicks with emulsifiers is favoured by the British authorities, this seems not to be true in other countries. Dispersion of the oil slick prevents beach pollution where the oil is more troublesome to deal with, and prevents further contamination of seabirds; furthermore, because the oil is now in the form of minute droplets it is argued that bacterial degradation of the oil is facilitated. The reason why dispersion is not more widely favoured is because of the fear that the oil is thereby widely distributed in the sea in a form in which it can be readily ingested by marine organisms and hydrocarbons enter the food chains with unknown consequences which are generally presumed to be undesirable. An alternative treatment, favoured by the French authorities at the time of the *Torrey Canyon* disaster, is to sink the oil by spraying it with a hydrophobic oleophilic powder. There has been some doubt about the fate of the sunken oil, which may return to the surface at a later stage, and fears that unless it remains immobilised in known areas, it is liable to foul fishing gear to the detriment of the fishing industry. More recently there has been a renewed interest in sinking agents in Britain (Zuckermann, 1969) and a variety of materials are being investigated in this country (Wardley Smith, 1968).

One outcome of the *Torrey Canyon* disaster was the discovery of the ease with which water-in-oil emulsions are formed at sea. The emulsion (known as "chocolate mousse") may contain 70–80 per cent water and is extremely stable. It has the physical properties of a gel rather than an oil, is not easily removed by emulsifiers and is less subject than oil to bacterial degradation (Dean, 1968). It follows that if oil slicks are to be treated with dispersants it is important to begin treatment as quickly as possible. Dispersion of an oil slick with emulsifiers is in any case most efficacious when the oil is fresh. Sinking agents are most effective with weathered oil. In either event the oil is transferred from the surface into the marine environment and it may be noted that besides being more widely distributed and more accessible to marine animals, dispersed fresh oil may be expected to contain a higher proportion of its lighter toxic constituents than weathered oil.

It is concern at the ignorance of the biological consequences of dispersing oil in the marine environment that has led North American authorities to favour physical containment and removal of oil wherever possible. Unfortunately booms have only a limited effectiveness in anything but quiet water and the various methods of removing oil from the water suffer the same disadvantage, though with the development of larger scavenging vessels this may be less of a problem of the future. Where special appliances of this sort are required, it is difficult and prohibitively expensive to mount an effective oil cleaning operation at short notice and at any point on the coast, so that it seems likely that physical removal of the oil, desirable as it may be, will be confined to a few sensitive areas.

Burning off the oil while it is still floating on the sea was attempted in the *Torrey Canyon* incident, with the aid of incendiary bombs and aviation spirit, but although it is estimated that some 20,000 tons of oil were destroyed in this way, this method of disposing of oil slicks was concluded to be uneconomic and relatively inefficient. The volatile and more inflammable fractions of crude oil were rapidly lost by evaporation and it was found difficult to sustain combustion because of the cooling action of the underlying water. In a collision of two tankers in the Baltic in March 1970 burning with the aid of a wicking agent had more success, even though the conditions in some respects were less favourable.

Special Problems of the Arctic

Most of our experience of oil pollution, its biological consequences, and the remedial measures that can be used to deal with it, has been gained in temperate and sub-tropical waters. The current development of the North Alaskan oilfield and the exploitation of the oil reserves that we may expect in the neighbouring Canadian Arctic introduce new problems for which previous experience outside the Arctic is not likely to be very helpful.

Because of the severity of the climate, operations connected with winning the oil and transporting it to the consumer are more difficult and more risky in the Arctic than elsewhere. To some extent the technical difficulties of exploiting the Arctic oilfields will oblige the oil companies to take extra precautions against accidents, but it is impossible to guard against all the dangers that may arise, and only prudent to assume that accidental oil spills will be an inevitable concomitant of exploitation.

The most serious risk of damaging pollution lies in the transportation of the oil from the oilfield to the main distribution centres; it is not necessarily a more hazardous operation than drilling, but it places very wide areas at risk. So far as the oilfields on the North Slope of Alaska are concerned, two methods of distribution are envisaged: by the Trans-Alaska Pipeline System, and by tanker through the north-west passage to the Atlantic ports of Canada and the United States. The latter route is being pioneered by the specially modified tanker *Manhattan* (106,500 tons d.w.) which completed the first return voyage in ballast from Halifax, Nova Scotia, to Prudhoe Bay in the late summer of 1969 and is repeating at least part of the voyage in the early summer of 1970 when ice conditions are more severe than before. A Canadian pipeline is also a possibility, particularly if oil in commercial quantities is extracted in the Tuktoyaktuk Peninsula. This would run up the Mackenzie valley to Edmonton, Alberta. An extension of this pipeline to Prudhoe Bay would provide an additional outlet for Alaskan oil; its route would traverse the drainage areas of a number of rivers: the Firth, Babbage, Blow, Rat and Ramparts.

The pipelines will be laid, at least in part, underground, and in order to pump the oil it will be necessary to maintain it at a temperature of 60–80 °C. The pipe will, of course, be insulated but even so there is no way in which some melting of the permafrost can be avoided. How extensive this will be cannot be predicted with certainty, but it is likely to cause local irreversible environmental damage. A more important danger is that the resulting differential settlement of the pipeline will fracture it and cause heavy oil pollution. If this occurs in the drainage areas of rivers the oil will be spread over a wide area. A further hazard exists where the pipeline runs in the beds of rivers or arms of the sea where it may be fractured by moving ice.

The Arctic sea route exposes an enormous length of often remote and inaccessible coastline to oil pollution, chiefly from wrecks and other accidents at sea. It is a hazardous sea route subject to violent storms, is inadequately charted and

66

with a constant threat from sea ice. On her maiden voyage, the *Manhattan* had some experience of these hazards: she was trapped in the ice for some days, passed over an abrupt and unexpected shoal near the Mackenzie delta with only 4·6 m clearance, and had a large hole ripped through her unprotected plates. The *Manhattan* is not a large tanker by modern standards and it is likely that economies of size will enforce the use of larger tankers if the Arctic route is opened. If there is sufficient traffic, a channel can be maintained ice-free throughout the year and, indeed, this route would be of little use to the industry unless it could be used on a year-round basis. Once opened, it would be used by a wide variety of shipping because it would reduce the voyage from the north-eastern United States and Europe to Japan and the Far East by a considerable fraction. This remote area may thus be expected to become one of the heavily trafficked sea-lanes of the world and, in view of the navigational hazards, it will inevitably be exposed to considerable pollution.

Because of its remoteness and inaccessibility, and the scarcity of harbours on the north coast, effective salvage and remedial operations will be difficult to mount. The problem is compounded by the lack of experience and technology for treating oil spills at arctic temperatures, and by the extreme sensitivity of the Arctic biological environment to disturbance of any kind.

At low temperatures, evaporation of the lighter and generally more toxic fractions of crude oil will be slow and the oil will remain in congealed masses. After the stranding of the *Arrow* in Chedabucto Bay, Nova Scotia, in February 1970, it proved impossible to pump oil from the stranded vessel into another tanker until it had been made less viscous by piping steam into the tanks. The low toxicity dispersant, Corexit 8666, designed for use at low temperatures, proved disappointing on that occasion and, so far, only the highly toxic oil spill removers seem at all effective for use in arctic conditions, and even their effectiveness on thick layers of congealed, viscous oil is much less than on oil spilled in more temperate waters. It may therefore be necessary to turn to alternative methods of treating oil spills as, for example, the use of wicking agents and burning off the oil.

Whatever remedial measures are used, it is likely that a considerable volume of toxic material will be added to arctic waters. Biodegradation, which is slow in temperate waters, will be slower still at arctic temperatures so that the consequences of an oil spill are likely to be much more persistent in the Arctic than in lower latitudes.

Hitherto, the general experience of oil spills in temperate waters has been that while some refined products may be very damaging, the long-term effects of crude oil pollution are slight, except to certain seabirds. Floating or dispersed oil does no material harm to plankton and although stranded oil and, still more, toxic oil spill removers may destroy much of the fauna and flora of beaches, recovery of these areas is relatively quick. Except for migratory and mobile animals, recolonisation of damaged areas is by the invasion and settlement of the dispersive phases (generally larvae) of the life histories of animals and plants which, as adults, live attached to the substratum or are otherwise static. Recolonisation by this means is an essential precursor to re-population of the area by the more mobile adult forms.

67

For reasons connected with the brevity of the breeding season and the uncertainty of the food supply, dispersive phases are rare among Arctic marine invertebrates; only about 5 per cent of Arctic species produce larvae which drift and feed in the plankton for an appreciable time and are therefore available to colonise new areas (Thorsen, 1936, 1950). The great majority of species produce small numbers of large eggs which are brooded by the parent and hatch as miniature adults that remain near the parent animals. This pattern of reproduction dispenses with the advantages of dispersion and exploration of new environments but there is evidence that it is much more economical of limited food resources (Chia, 1970). This very low reproductive rate is further reduced by the fact that because of the extremely short summers, Arctic marine invertebrates produce only one crop of young each year (and sometimes not even that) whereas their close relatives living in temperate waters may breed several times a year.

Coupled with a low annual reproduction, polar invertebrates are generally extremely slow growing, take several years to attain sexual maturity, and may live to a very great age. The best documented study of this is by Pearse (1969) of the Antarctic starfish, *Odonaster validus*, which increases in weight only 1–2 grams per year, may live as long as 100 years and takes 3–6 years to reach sexual maturity. A major pollution incident which destroyed an animal community made up largely of species with a characteristically "arctic" life history, would eliminate not only the adults but also the young they were brooding. Replacement of them from adjacent areas would be extremely slow because of the lack of mobile phases in the life history of most species, and the re-establishment of a population with a balanced age structure might take many decades.

This type of reproductive biology with a low reproductive rate and long life is a feature not only of marine invertebrates but of a large proportion of plants and animals in the Arctic. Recovery from additional mortality caused by pollution will be correspondingly slow for all of them. It is significant that the seabirds which suffer most dramatically from oil pollution in temperature waters are precisely those species which are essentially Arctic birds with a reproductive biology characteristic of Arctic animals. An additional hazard to seabirds in the high Arctic is their tendency to be concentrated in enormous flocks on limited areas of open water. Particularly during moulting and in the pre-migratory flocking, a substantial fraction of the world population of some species may be concentrated in a single bay. A serious oil spill at a critical time might easily have long-term repercussions on a species out of all proportion to its size and on a scale completely unprecedented in more southerly waters.

Public Health Hazard

At present it is impossible to say whether or not oil pollution presents a hazard to human health, but there is a certain amount of information that suggests it would be wise to give this matter urgent attention so that a proper assessment can be made.

Whatever hazard there may be resides chiefly in the carcinogenic polycyclic hydrocarbons, including 3,4-benzpyrene. Although these substances are present in mineral oils, they also occur naturally in plants, the soil, fresh waters and, as a result of combustion of organic matter, in air polluted by smoke and diesel and petrol fumes. These substances are therefore part of our normal environment. Their concentration in potable water does not usually exceed 0·025–0·100 μg/l (Borneff & Kunte, 1965), and since threshold concentrations for most petroleum products at which they can be detected by humans are at or below this figure (Ineson & Packham, 1967), seriously polluted waters are unlikely to be drunk in quantity because of their taste. It is recommended in a WHO publication (Suess, 1968) that the polycyclic hydrocarbon content of water supplies should not exceed one-tenth the amount taken up from the atmosphere of a city and it is unlikely that drinking water constitutes a public health hazard because of contamination by oil or petroleum products (Martin, 1970).

The sensitivity of human beings and their revulsion to the taste of mineral oils is also likely to provide an effective barrier to the consumption of oil or refined products by eating contaminated fish or shellfish. Indeed, the economic damage to fisheries which is caused by contaminated fish reaching the market, or even the suspicion that they may reach the market, is sufficient evidence that there is little danger from that source. Breakdown products or degradation products of mineral oils are not so easily detected, however, and it is these which may present a hazard.

Hydrocarbons, whatever their structure, tend to be stable once they have been incorporated into a marine organism and can pass through a number of links in a marine food chain without alteration. The stability of hydrocarbons is so great that hydrocarbon analysis has been used as a means of studying food sources (Blumer *et al.*, 1970).

A considerable quantity of oil is legally and deliberately discharged at sea and under the new proposed convention it will be finely dispersed rather than concentrated into oil slicks. Because of its dispersion, it will be the more readily ingested by marine organisms. Parker, Freegarde & Hatchard (1970) have shown that fine droplets of crude oil are ingested by the zooplanktonic organisms *Calanus* (a copepod) and nauplius larvae of *Balanus* (a barnacle). The oil appears, apparently unchanged, in the faeces of these animals but has already entered the food chain, being transmitted to organisms that feed on the zooplankton or detritus feeders who consume the oil-laden faeces. Even if *Calanus* and *Balanus* nauplii are unable to metabolise the oil, it cannot be expected that other animals also cannot do so, or that its breakdown products resulting from bacterial activity will not enter the food chains.

Carcinomas and papillomas have been found on the lips of a Pacific food-fish, the croaker, in an area polluted by oil refinery wastes (Russell & Kotin, 1956) and papillomas have now been observed on the lips and body of a number of food-fishes in polluted areas (Young, 1964). There is no evidence to suggest that these cancerous or precancerous conditions are transmissible to human beings, nor is there evidence that carcinogenic hydrocarbons derived from oil or petroleum products accumulate in organisms to the extent that they constitute a real hazard to human health. Equally, there is no evidence that they do not. In fact, there is remarkably little evidence of any sort and, furthermore, it would be hard to get. Severe and prolonged exposure to carcinogenic compounds is necessary to produce cancers in humans and the effect is generally long-delayed, so that it is difficult to identify a causative agent. The effect, even if it can be reasonably established, is only statistical and for this reason it is easy to dismiss the dangers as trifling or imaginary. Nevertheless, now that it is clear that there may theoretically be a hazard and that there is evidence, even though it is only suggestive, that some hazard may actually exist, it would be foolhardy not to investigate the matter further.

International Conventions and Control Measures

Sporadic pollution by crude oil in the form of oil slicks is damaging to certain seabirds, is a nuisance on tourist beaches and may cause temporary damage to the flora and fauna of the shores and, less directly, to fixed coastal fisheries. Otherwise, it is not apparently harmful. The obvious treatment hitherto has been to disperse the slicks with oil spill removers, but because until recently only toxic dispersants were available and these were likely to be more damaging than the oil unless used with care, other treatments have sometimes been preferred. These include sinking the oil (although this introduces the chance that the sunken oil will contaminate fishing gear and taint the catch) and using various means to remove it physically (although these are practicable only for relatively small oil spills in sheltered water).

These conclusions must be qualified in a number of ways. First, chronic pollution is always a serious problem for it does not permit natural recovery from damage. The cumulative effect is the progressive impoverishment of the natural resources of an area. Second, these remarks apply to crude oils, most heavy refined products and heavy residues. Light refined petroleum products are much more toxic and if spills occur close inshore or in relatively enclosed waters their effect is so swift and damaging to all marine life that effective remedial action is difficult. Unfortunately, much of the traffic in these products is coastal and in rivers, and the risk of accidental spills and the likelihood of extensive damage are both much greater than for seaborne crude oil. There seems little that can be done except to treat refined petroleum products as dangerous cargoes and handle them with appropriate care, although this is hardly a very satisfactory solution. Finally, certain environments are more vulnerable than others to pollution and require special measures to safeguard them or to treat oil pollution when it occurs there. Nursery and breeding grounds of commercial fishes are one category of such environments. The Arctic, because of its excessively slow recovery from damage, is likely to be another. Fears have been expressed by Australian conservationists about the threat to the Great Barrier Reef by exploitation of the offshore oilfield in Queensland. The Barrier Reef is certainly a unique and complicated biological formation but whether or not offshore drilling will constitute a real threat to it is disputed and the conflicting claims are at present being examined by a Royal Commission.

International agreements are designed not to legislate for appropriate treatment of pollution but to prevent it at source. So far as oil pollution at sea is concerned, by far the greater amount has been, in fact, deliberate and avoidable and has been caused by the discharge of dirty ballast water, tanker washings, bilge water and the like. While some steps have been taken and more are being discussed to regulate shipping in heavily trafficked sea-lanes to reduce the risk of collision, international agreements in the last two decades have been aimed at reducing and, if possible, prohibiting avoidable discharge of oil, particularly in coastal waters.

The 1954 and 1962 international conventions did much to reduce the level of deliberate oil pollution and the introduction of the "load on top" system of retaining tanker washings even more. However, the latter is in technical breach of the earlier conventions and a new agreement proposed by IMCO effectively legalises it by eliminating the present limit of 100 mg/l oil in water for permissible discharge and replaces it by a rate of 60 l/mile of ship travel as the legal limit for the discharge of oil.

This proposal is concerned principally, though not entirely, with the discharge of crude oil by tankers. Crude oil causes immediate and obvious damage when it is in the form of a slick. Since the new code of practice would avoid the formation of oil slicks, the new proposals are sensibly designed to deal with a major part of the problem of oil pollution.

This being so, it may seem ungenerous to criticise other aspects of the new IMCO proposals. The earlier proposals aimed at the ultimate prohibition of deliberate discharge of oily waste and included the stipulation that all new tankers over 20,000 tons should be equipped to retain all waste oil. This is no longer required. Second, in the new proposals, the coastal zone in which any discharge of oil is prohibited is limited to 50 miles. The reasoning underlying these changes is the supposition so long as oil does not form slicks it is not damaging. This may be true so far as present certain knowledge is concerned, but the oil, although not visible, will still be introduced into the marine environment and, indeed, more widely distributed than before. What is more, with the narrow prohibited coastal zone, there will be an increased release of oil in areas on the continental shelf where there is the greatest productivity and the principal stock of commercial fish.

In matters such as environmental pollution it is not safe to wait for conclusive proof of damage before acting to control it. Disturbance of biological systems has widespread repercussions and they are generally slow to manifest their full effects. By the time damage has been demonstrated to everyone's satisfaction and control measures instituted, the damage may well be irreversible on any reasonable human time scale. This is a situation in which legislation must often be in advance of firm scientific knowledge and great attention must be paid to the first tentative evidence that a dangerous situation may exist.

Evidence that low-level oil pollution has long-term consequences is at present slight. The fact that it exists at all should be a matter for concern and immediate investigation. It would be unrealistic to expect industry to respond to the first wisp of evidence, but it would be folly to assume that by attempting to reduce the number of oil slicks that are formed we shall have solved the problem of oil pollution. There is a clear danger that having made the pollutant invisible and reduced its more obvious damaging effects, the possibility that it may have more subtle, long-term consequences will be ignored.

71

References

BAKER, J. M. (1970). *Mar. Polln Bull.*, **1** (2), 27–8.
BEER, J. V. (1968a). *Field Studies*, **2** (suppl.), 123–9.
BEER, J. V. (1968b). *Wildfowl*, **19**, 120–4.
BEER, J. V. (1970). *Mar. Polln Bull.*, **1** (6), 84–5.
BERGMANN, G. (1961). *Suomen Riista*, **14**, 69–74.
BLUMER, M. (1969). *Oceanus*, **15** (2), 2–7.
BLUMER, M., SOUZA, G., & SASS, J. (1970). *Mar. Biol.*, **5**, 195–202.
BORNEFF, J., & KUNTE, H. (1965). *Arch. Hyg. Bakteriol.*, **148**, 583.
BOURNE, W. R. P. (1968). *Field Studies*, **2** (suppl.), 99–121.
CHIA, F. S. (1970). *Mar. Polln Bull.*, **1** (5), 78–9.
CHIPMAN, W., & GALTSOFF, P. S. (1949). *U.S. Fish Wildl. Serv. Spec. sci. Rep.*, **1**, 1–53.
CLARK, R. B. (1968). *Lancet*, Oct. 5, 1968, 770–2.
CLARK, R. B. (1969). *Proc. intern. Conf. Oil Pollution of the Sea*, Rome, 1968, 76–112.
COLE, H. A. (1969). *Oceanology International*, 1969.
DARLING, F. F. (1938). *Bird flocks and the breeding cycle* (Univ. Press, Cambridge).
DEAN, R. A. (1968). *Field Studies*, **2** (suppl.), 1–6.
GILES, L. A., & LIVINGSTON, J. (1960). *Trans. N. Amer. Wildl. Nat. Res. Conf.*, **25**, 297–302.
GROSS, A. C. (1950). *Proc. X intern. Congr. Ornithol.*, 532–6.
HAMPSON, G. R., & SANDERS, H. L. (1969). *Oceanus*, **15** (2), 8–10.
HARTUNG, R. (1963). *Pap. Mich. Acad. Sci., Arts, Letters*, **48**, 49–55.
HARTUNG, R. (1965). *J. Wildl. Mgmt.*, **29**, 872–4.
HARTUNG, R. (1967). *J. Wildl. Mgmt.*, **31**, 798–804.
HARTUNG, R., & HUNT, G. S. (1966). *J. Wildl. Mgmt.*, **30**, 564–9.
HAWKES, A. L. (1961). *Trans. N. Amer. Wildl. Nat. Res. Conf.*, **26**, 343–55.
INESON, J., & PACKHAM, R. F. (1967). Joint problems of the oil and water industries (ed. P. Hepple), 97–107. (Inst. of Petroleum, London.)
KASYMOV, A. G. (1970). *Mar. Polln Bull.*, **1** (7).
KORRINGA, P. (1968). *Helgol. wiss. Meeresunters.*, **17**, 126–40.
LAEVASTU, T. (1960). FAO Fisheries Division Project (FAO, Rome), No. 2 (21/3).
LEMMETYINEN, R. (1966). *Suomen Riista*, **19**, 63–71.
MACKIN, J. G., & SPARKS, A. K. (1962). *Publ. Inst. Mar. Sci. Texas*, **7**, 231–61.
MCCARTAN, L. (1958). *Brit. Birds*, **51**, 253–66.
MARTIN, A. E. (1970). *Symp. Water Pollution by Oil*, Aviemore, 1970, pap. 13, 1–7.
MILON, M. (1966). *Terre et vie*, **20**, 113–42.
NELSON-SMITH, A. (1968). *Field Studies*, **2** (suppl.), 73–80.
NELSON-SMITH, A. (1970). *Symp. Water Pollution by Oil*, Aviemore, 1970, pap. 23, 1–8.
NORTH, W. J. (1964). *Symp. Poll. mar. Micro-org. Prod. pétrol*, Monaco, 1964, 335–54.
O'CONNOR, R. J. (1967). *Seabird Bull.*, **4**, 38–45.
PARKER, C. A., FREEGARDE, M., & HATCHARD, C. G. (1970). *Symp. Water Pollution by Oil*, Aviemore, 1970, pap. 17, 1–8.
PARSLOW, J. L. F. (1967a). *Brit. Birds*, **60**, 1–41, 177–202.
PARSLOW, J. L. F. (1967b). *BTO News*, **23**, 8–9.
PEARSE, J. S. (1969). *Mar. Biol.*, **3**, 110–6.
RITTINGHAUS, H. (1956). *Orn. Mitt.*, **8**, 43–6.
ROSKAM, R. T. (1965). CM Council Meeting, Int. Council Explor. Sea, Sect. C, Northern North Sea Committee, 44.
ROSKAM, R. T. (1966). *Wat. Bodem Lucht*, **56**, 19–21.
RUSSELL, F. E., & KOTIN, P. (1956). *J. Nat. Cancer Inst.*, **18**, 857–61.
SELECT COMMITTEE (1968). Coastal Pollution. *Rept Select Comm. Sci. Technol.*, House of Commons, 421-1 (HMSO, London).
SIMPSON, A. C. (1968). *Field Studies*, **2** (suppl.), 91–8.
SMITH, J. E. (ed.) (1968). *"Torrey Canyon" Pollution and Marine Life* (Univ. Press, Cambridge).
SOUTHERN, H. N., CARRICK, R., & POTTER, W. G. (1965). *J. anim. Ecol.*, **34**, 649–65.
STRAUGHAN, D. (1970). *Mar. Polln Bull.*, **1** (4), 61–2.
STRAUGHAN, D., & ABBOTT, B. C. (1970). *Symp. Water Pollution by Oil*, Aviemore, 1970, pap. 21, 1–3.
SUESS, M. J. (1968). *Polynuclear aromatic hydrocarbons*. WHO/W Poll/68.4. (WHO, Geneva.)
TANING, J. J. C. (1952). *Sveriges Nat.*, **5**, 114–22.
TARZWELL, C. M. (1970). *Symp. Water Pollution by Oil*, Aviemore, 1970, pap. 22, 1–10.
TATE, C. (1935). *Brit. Birds*, **28**, 314.

THORSEN, G. (1936). *Medd. Grønl.*, **100** (6), 1–155.
THORSEN, G. (1950). *Biol. Rev.*, **25**, 1–45.
TUCK, L. M. (1960). *The murres: their distribution, population and biology, a study of the genus* Uria. (Wildlife Service, Ottawa).
USPENSKII, S. M. (1956). *Bird bazaars of Novaya Zemlya* (Acad. Sci. USSR, Moscow).
WARDLEY SMITH, J. (1968). *Field Studies*, **2** (suppl.), 15–19.
WIEBE, A. H. (1935). *Trans. Amer. Fish. Soc.*, **65**, 324–31.
WOOD, P. C. (1968). *Nature, Lond.*, **220**, 21–2.
YOUNG, P. C. (1964). *Calif. Fish and Game*, **50**, 33–41.
ZUCKERMANN, S. (1969). *Proc. intern. Conf. Oil Pollution of the Sea*, Rome, 1968, 148–59.

Oil Pollution: Industry

MR MAURICE HOLDSWORTH (Shell Petroleum)

When I received the papers for this conference like most of you, I imagine, I was very interested to see who was speaking. This Panel seemed to be slightly differently constituted from the rest. We have a chairman, we have a lawyer, we have a scientist and a so-called administrator, but I noticed that the scientist is a very well-known and established conservationist and, of course, the so-called administrator is a man from an oil company.

Some of you, if you reacted in the same way as I did, may have seen a situation where, perhaps, the scientist-conservationist was going to have a bash at industry and vice versa. I have in my notes: "Well, of course, this couldn't happen here in such a well-informed audience and panel as this." However, having heard the last few remarks of Professor Clark, I feel there is still room for a little more communication in this field between industry and science.

Thinking thus of the relationship of Industry and Conservation, it seemed I might usefully take as my theme, "The Place of Industry in the Community's Environment".

Conflicts between conservationists and industry do, of course, happen. The conservationist often says that industry is unwilling to do anything that will reduce its profits because the only concern of industry is to maximise profits for its shareholders. Industry is largely to blame for this attitude arising because at any business school this tenet is constantly drummed in. However, the profit motive of industry is a small part of the chain of the total mechanism linking industry and the community. In the communist countries, the direct profit motive is not there, but I am quite sure that industry there has very much the same problems of communication and decision in this matter of conservation as we have in our free enterprise capitalist society.

Industry, surely, can only exist to provide the community with the services that the community desires. Industry is not a separate entity which is somewhere apart generating its own activity and, as it were, the enemy to be attacked and controlled. Industry is part of the community, it exists for the community, and it is the community's right, not merely privilege, to require from industry what the community wants. To base the issue on the day to day interplay of profits and costs is, to my mind, irrelevant in the long term.

73

This morning, I think it was Professor Burhop who said: "Industry must be told that these costs must be found." Well, I would not in any way cavil about the content of that, but I rather suspected that the phraseology implied that the costs could be found from some secret coffers. Surely, there is only one place the costs can come from and that is from the community. The community puts down what it wants, when it knows what it wants, and industry must comply and produce the service to that standard.

Now, of course, the interaction between industry and the community has gone on for a long time, and this control by the community is inherent and inevitable. It is a process of demand and then acceptance or rejection of the service which is offered. This process at times can be a rather tedious matter of trial and error and, perhaps, as Professor Clark says, sometimes all too much error.

In engineering terms one might say that this "control loop" in the sphere of environmental conservation has in the past been a rather loose undamped one with minimum response, but it has probably been good enough for the rate of expansion of industrialisation and of the community in earlier times. However, we are advancing now so rapidly that we desperately do need a much tighter control loop and with – if I might be forgiven another engineering term – a derivative function so that we can detect the trends and control them before they become a real problem.

The modern rate of expansion is in no sphere more apparent than in the oil industry which is subject to a quite tremendous expansion. In 1950 there were 500 million tons of oil brought out of the ground and turned into energy products. In 1970 this has risen to 2,000 million tons. In two decades four times the size. When we come to oil tankers the growth in that particular section of the oil industry has been even more phenomenal. In 1950 there were about 80 million tons of crude oil transported by sea and in 1970 this will top 800 million tons. A ten-to-one increase in two decades. Thus, the problem is a very real one and we do need tighter control than the "laissez faire" system of interaction between community and industry which we have had in the past. This conference and the multiplicity of similar gatherings in this Conservation Year point up a growing awareness of this need. However, awareness alone will not solve the problem and we need now to put together the design of both national and international control loops. Most of the nuts and bolts are already available. The establishment of the United Nations Agencies has provided a most workable mechanism whereby the international legislation can be formulated by unemotional, factual dialogue between industry and the community and sensible solutions arrived at.

I would like to take now three examples of where I see the mechanism already beginning to work. First the case of the Bavarian refineries. Until recently oil products in Bavaria were brought to Bavaria by means of road and rail transport across Europe from where they were refined on the Channel and the North Sea coast. This was expensive, and as a result the cost of energy from oil in Bavaria was some 50 per cent greater than in the rest of Germany. The new concept, which is now working, was to bring the crude oil in pipelines from the sea-board to Bavaria and there refine it on the spot. This has brought down the finished cost by almost a third. Bavaria now enjoys all the benefits of substantially reduced cost of energy which, as we have heard from various speakers this morning, can

74

have quite a chain effect in the whole economy of the region. Part of the total available saving was used for building into the refineries the highest degree of effluent and environment control. Before such industry could be brought into this unspoilt area of natural beauty, clean air and unpolluted water, there was full discussion and agreement between the oil companies and the community. The community wanted the economy of the refineries and were prepared to forgo part of the full gain in order to have the minimum interference with the environment. The environment has been changed but not significantly deteriorated.

My second example has already been touched on – the International Convention on the Prevention of Pollution of the Sea by Oil. The voice of the community was probably first heard through the Advisory Committee on the Prevention of Pollution of the Sea in the early 'fifties when the problem really began to make itself felt. The problem arose after the Second World War because of the rapid increase in crude oil transportation which came about because, like the Bavarian refineries, refineries were erected in the consuming areas of north-west Europe rather than, as formerly, in the Middle East where the oil came out of the ground. This meant that, instead of carrying refined products around the world, we now started to carry crude oil and in ever-increasing quantities to meet the demand. A crude oil tanker must wash out its tanks some time on its ballast voyage; in those early days all the oil clinging to the inside of the tanker was washed into the sea, and the pollution of our beaches began to be intolerable. As a result of the voice of the Advisory Committee – and I think it is fair to say they were the prime movers in this – the 1954 Convention came into being. It was a well-intentioned Convention but it has been far from fully effective. The present pollution on our beaches is not very different from that of 1950. This is compounded of two things; the enormous increase of oil carriage, as we have seen, coupled with a large reduction in the amount of oil per unit of oil carriage which now is put into the sea. However, the end result of the high figure multiplied by the lower factor is still pollution. One reason for this was that not enough was known by industry at the time of the 1954 Convention to see a sensible way of solving the problem and the interim measure proposed by the Convention was to concentrate the dirty washings and dump it at least 100 miles from any coast. Unfortunately, this oil is very persistent and sticks together in little balls and much is eventually driven on to somebody's shore despite the 100 miles offshore requirement.

The oil industry in the late 'fifties saw that this just was not working and proposed the load-on-top system, which can reduce the effluent from a ship from this very large amount, measured in hundreds of tons, to a very small amount. Up to now the operation of load-on-top has been voluntary. Although the system reduced tremendously the amount of oil put into the sea from a ship, it could not meet the Convention's effluent requirements of less than 100 parts per million of oil at all times. Secondly, being a voluntary agreement, and this being completely an international problem, it was not possible to persuade everybody in the worldwide oil industry to use the system. The system, in many cases, is costly and the difficulty here is that the cost is not on the ship which is doing the polluting but in the refinery eventually receiving these residues which otherwise

75

would have been put into the sea. Although there is generally in the oil industry a fairly cohesive interchange between the various functions, nevertheless, it was not possible to persuade all refiners throughout the world, many of them government owned refiners, to accept the financial onus for what they regarded as somebody else's problem, producing pollution hundreds of miles from their own locality. Some 80 per cent of refineries throughout the world responded but the remainder still find themselves unable to co-operate. This situation has now been explained to the community via the channels of IMCO, who now have the Convention in their hands. New amendments to the Convention are coming forward which will, if they are enforced, reduce this pollution down to what I hope will be an insignificant amount. Just to give an idea of the scale of the problem: if nothing was done and at the present rate of crude oil transport we continued in the same way as was going on in the early 'fifties, then 3 million tons of oil would go into the sea per year by this particular action. As a result of load-on-top, which is applied to only 80 per cent of ships, there is still going into the sea, from the other 20 per cent, probably something like 600,000 tons of oil per year. With the introduction of the new amendments and their enforcement, the total oil worldwide from this particular source should be cut down to something between 15 and 20,000 tons of oil per year. Of course, 15 to 20,000 tons of oil dumped anywhere in one lump is an awful lot of oil, but we must remember that this is the total amount of oil going into the sea worldwide over thousands and thousands of miles of ocean and the dilution factor then is really something. I hesitate to bring in these sorts of trends because, for too long, we have regarded the sea as being so great that it will accept and absorb any amount of contaminent. Of course this is not true, but this reduction from 3 million potential tons of pollutant down to 20,000 tons is an enormous step forward.

My third example of industry being informed of what the community's will is and taking action to meet it concerns marine pollution accidents. First I should put this in perspective. The pollution we should be most concerned with is chronic pollution that goes on, day by day, by deliberate discharges in operations. Of course, when there are accidents like the *Torrey Canyon*, the impact on the community is most newsworthy and, of course, it is a dreadful disastrous impact. The resultant pollution, however, is confined to a locality and it does not greatly affect the total ecology. Three million tons per year, for instance, going into the sea would be equivalent to a *Torrey Canyon* every ten days or so. The 600,000 tons per year still going into the sea is equivalent to a *Torrey Canyon* every two or three months. Nevertheless, accidents do impinge disastrously on a locality and have in the past put innocent sufferers to great expense in clearing up the mess. The community quite rightly has said that this is not fair and that those who carry oil and spill oil ought to be the ones who pay for clean-up. Maritime law, at the moment, is not at all clear on this and more appropriate law is still to be forged. However, it is obviously fair that the locality whose beaches are suddenly inundated by oil from an accidental spillage should be supported by money immediately available from some source which is properly connected with those who spill the oil. Recognising this soon after the *Torrey Canyon*, industry brought about a voluntary insurance scheme which will recompense governments for major oil spills up to $10 million total. A recent diplomatic conference – voicing the will of the community – has shown that there is required an even greater and wider protection which will apply whether the ship that spills the oil is at fault or not. As a result of this, action, again on a voluntary basis by industry, is being taken to provide an international fund which will be available to complete the recompense wherever the previously mentioned insurance scheme falls short of what is required.

I hope I have shown that mechanisms exist and are growing whereby industry and the community can have a sensible dialogue to arrive quickly at practical solutions to environmental problems. It must be remembered, however, that industry is the servant of the community and as such has not the wherewithal to pay for improvements to the master's household. The servant, however, should and must be worthy of his hire.

Oil Pollution: the Law

PROFESSOR R. Y. JENNINGS (Professor of International Law, University of Cambridge)

I take it that my task is to suggest what contribution the law can make to this problem of oil pollution and also to summarise what has been accomplished so far.

It may be useful to begin with a statement of some of the difficulties that the lawyer encounters at the outset.

First, legal regulation of a matter of this sort involves both international law – mainly in the form of treaties or conventions – and the separate domestic laws of different states. A crucial question, therefore, is the method and procedures by which the international laws are translated into legal and administrative action in the separate states. These can be effective and they can sometimes be less than effective.

Secondly there is the immense complication that oil pollution affects and straddles several legally quite different zones of waters and of sea-bed and sub-soil. There are the national waters of a state, e.g. the waters of a port; there are the territorial waters with potential controversy about their proper breadth; and these waters are all, in somewhat different ways, subject to the sovereignty of the coastal state. But then there are the high seas, the common heritage of all and governed by the principle of freedoms enjoyed by all. And if this were not enough variation, there is the continental shelf underlying the high seas, on which the coastal state enjoys sovereign rights, the exercise of which may well pollute both high seas and territorial seas. And beyond the continental shelf – and what *is* "beyond" is itself a doubtful question – there is a regime governing the sea-bed and sub-soil which is again different but also embryonic and controversial.

It will immediately be clear that waters subject to a particular state's sovereignty cannot be subjected to an international scheme of regulation unless that state can be persuaded to subscribe to the treaty; but on the high seas the problem is even greater; for any qualification by treaty of the freedoms enjoyed

77

under *general* customary international law by *all* states will need the subscription of most maritime states if it is to be really effective. It is in fact a somewhat extreme example of the familiar difficulty that there is no developed procedure of legislation in international law, by which I mean some way of making a law which will bind even an opposing minority of states.

Thirdly, there is the difficulty that the traditional mode of international law is to provide a remedy in damages ("reparation") for an injury, provided there is a state qualified and willing to pursue the claim. But this notion of a remedy is not entirely helpful in dealing with a problem that is basically one of public order and social regulation rather than one of monetary reparation for injury already suffered.

Fourthly, the problem itself is extremely complex, for usually there is no obvious malefactor whose activities need to be subject to a simple prohibition. Pollution results from lawful, salutary and even necessary activities such as getting the resources of the sea and sea-bed, the transport of fuels and even ordinary navigation. So there is no simple solution even in theory; quite apart from the difficulties of translating a proposed solution into effective law.

Yet in spite of the difficulties it is the fact that more has been done about oil pollution at sea than about any other major kind of pollution. The reason I suspect is just that you can actually *see* oil pollution. It is visual pollution as well as chemical or biological pollution. And then there was the *Torrey Canyon* disaster which caught the public imagination very strongly; though it is perhaps doubtful how far it would have resulted in such rapid action had it not been for the thoughtful initiatives of the British Government, and it is proper here to pay tribute to the civil servants who did the necessary study for the groundwork of action, quietly, with virtually no publicity and little thanks.

But, as we shall see in a moment, almost all the progress has been made in regard to the *Torrey Canyon* kind of situation and still almost nothing has been done about the dangers from offshore oil drilling; though the Santa Barbara Channel disaster showed that it is no less great.

Yet although the *Torrey Canyon* case lent great impetus to the endeavour to control oil pollution at sea it must not be supposed that international lawyers at least were taken by surprise. It was in 1937 that the Institute of International Law called upon governments to study ways and means of controlling pollution at sea, adding moreover that the failure to do so would be a breach of international responsibility.

There are clearly three main lines of legal endeavour required:

    (i) rules to prevent pollution from happening;
    (ii) rules and procedures to contain and mitigate a disaster that has already begun to develop;
    (iii) procedures and rules to provide a remedy in damages for injury resulting, including the cost of containment and mitigation measures.

Such international legislation as we now have tackled those three problems in that logical order.

Prevention was the object of the 1954 London *Convention for the Prevention of the Pollution of the Sea by Oil*; of course, the ambitious title is explained by the circumstance that in 1954 it was possible to believe that, if you could control discharge of oil from tankers, you had to all intents and purposes dealt with the problem. The Convention entered into force for the parties on 26 July 1958; and it was amended and strengthened in important respects in 1962 and again in 1969. It is still the only piece of comprehensive international legislation that is aimed at preventing pollution before it occurs.

The principle of the Convention is to lay down fairly stringent conditions, under which alone, oil or oily mixtures, may be discharged from ships into the sea. The details are highly technical and do not matter for our present purpose. A form of "Oil Record Book" is laid down in a schedule. And there is a system of inspection by any party to the treaty. The means of enforcement are quite impressive, at any rate on paper. Yet one gathers that detection of offences is in reality far from easy. The penalty for offences depends very much on the flag state. And of course this prevention convention does not touch the *accidental* oil spill of the *Torrey Canyon* kind. Nor does it touch oil seepage whether from drilling or from natural causes. And even in those aspects of oil pollution that it does deal with, the aim is the protection of coasts and coastal waters rather than the high seas itself. Moreover, recent as the latest version of this Convention is, we may soon be faced with a new set of problems arising from new technology; for the Japanese are considering the construction of large tankers which would refine the oil during the voyage; what is to happen to the waste products, considering, as I am told is the case, that one of the purposes of these refinery ships would be to reduce the pollution which results from refineries on land?

The Brussels Conference of 1969, held under the auspices of the IMCO, and at the initiative of the British Government, produced two new conventions dealing respectively with (i) intervention to deal with accidents that have already occurred, and (ii) liability for damage.

The 1969 *Convention Relating to Intervention on the High Seas in cases of Oil Pollution Casualties*, or, for short, *The Public Law Convention*, gives a party, as against other parties, a right to go out on to the high seas and take certain measures on the high seas to defend itself against advancing oil slicks. It is no doubt arguable that states have some such right anyway under some kind of law of self-defence. I seem to remember a letter to *The Times* to that effect from Lord Shawcross after the *Torrey Canyon* affair. But such rights even if they exist are most unclear and controversial and it is certainly desirable to spell it all out in treaty form, as is done in Article I:

"Article I:
"1. Parties to the present Convention may take such measures on the high seas as may be necessary to prevent, mitigate or eliminate grave and imminent danger to their coastline or related interests from pollution or threat of pollution of the sea by oil, following upon a maritime casualty or acts related to such a casualty, which may reasonably be expected to result in major harmful consequences."

And then a somewhat serious qualification of that:

"2. However, no measures shall be taken under the present Convention against any warship or other ship owned or operated by a State and used, for the time being, on government non-commercial service."

Article II defines a "maritime casualty" as "a collision of ships, stranding or other incident of navigation, or other occurrence on board a ship or external to it resulting in material damage or imminent threat of material damage to a ship or cargo".

Article III, however, provides certain requirements of consultation and notification for a state proposing to act under Article I. There is a provision for the establishment of a panel of "experts", who shall be consulted; and there are provisions for the settlement of disputes.

The procedures for consultation, notification and co-operation are carried a good deal further by North Sea countries in the Bonn Agreement of 1969 *Concerning the Pollution of the North Sea by oil*.It provides for standing machinery not only in fact in respect of the North Sea proper, but also of the English Channel including the western approaches as far out as the Scillies.

This precedent may be followed in other suitable geographical regions. Kuwait has recently issued invitations to the Persian Gulf States to a conference to produce an oil pollution treaty for the Gulf.

As has already been mentioned, the Brussels "private law" convention dealt with the question of civil liability, *viz*. who is to pay for it all? The dubiety of the answers to this seemingly simple question are readily illustrated from the *Torrey Canyon* case itself. The vessel was owned by the Barracuda Tanker Co. of Bermuda. But Barracuda was a subsidiary of Union Oil of California. For tax reasons the vessel had been leased by the subsidiary to the parent company, but she was registered in Monrovia and flew the Liberian flag. The crew was Greek, the captain Italian, and she was on charter to BP.

Thus, when the British Government sought to recover some damages, there was a choice of suing Union Oil in California, Barracuda in Bermuda, the vessel and its owners in Liberia, or any of these in the British courts. And of course this represents not only a variety of venues but also a variety of different laws of responsibility and reparation. After various sorties in two jurisdictions a settlement was reached under which the owners and charterers indemnified the British and French Governments to the tune of £1½ million apiece; as against the estimated £5 million the incident cost the British Government alone.

These then are the questions which the second Brussels Convention tackled. There were several difficult questions on which views both before and at the Conference differed considerably. For instance: should there be strict liability, or fault liability with the burden of proof reversed?; through whom should the liability be channelled, owner of the vessel, of the cargo, the operator, and so on (he would of course have his own rights of recourse against the others)?; should there be a scheme of compulsory insurance, or of some other guarantee of financial competence to meet a claim?; should there be limited liability?

These are all questions to which there is no right answer. The solution can only be a compromise and clearly a whole spectrum of compromises is possible on the various combinations of the various answers to these and other questions. Nevertheless a good Convention was produced by the Conference, liability being channelled through the registered owner of the vessel on the basis of a qualified form of strict liability designed to make the risk an insurable one. The Convention does not cover hostilities, or "natural phenomenon of an exceptional, inevitable and irresistible character"; which appears to be a pompous circumlocution for "Act of God".

This Convention, then, for the parties to it, provides reasonably neat machinery for recovering the cost, or some of it, of prevention and clean-up in case of maritime casualties. It does not, of course, do anything for the drilling casualty. Nor does it do anything for the high seas which are the common heritage of mankind. It is only "territory" of a party that is the object of the remedies provided.

The international treaty is not the only, nor indeed is it necessarily the most effective way of doing something about pollution by oil. Another possibility is for a state to persuade itself that it is legally justified in unilaterally extending its governmental pollution regulations beyond the limit of the territorial sea, or even the contiguous zone on to the high seas. The latest action on the pollution front is just that; the Canadian *Arctic Waters Pollution Prevention Act*, 1970. This comprehensive and vigorous piece of legislation is said to be justified (and perhaps it is) by the peculiarly damaging effect of oil pollution in cold waters.

This Act is intended to deal with all kinds of oil pollution, whether from ships or from operations on the sea-bed; and it provides for measures both of prevention, of intervention after an escape of oil and for sanctions.

It begins with a preamble which sets out the purpose of the Act, i.e., to implement the obligation of Parliament to ensure that Canadian Arctic resources are exploited and that Canadian Arctic waters are navigated "only in a manner that takes cognizance of Canada's responsibility for the welfare of the Eskimo and other inhabitants of the Canadian Arctic and the preservation of the peculiar ecological balance that now exists in the water, ice and land areas of the Canadian Arctic".

"Arctic waters" are then defined for the purposes of the Act. They are waters contained within the 60th parallel of north latitude, the 141st meridian of longitude on the west, *viz.* the longitude which proceeds through the boundary with Alaska, and "a line measured seaward from the nearest Canadian land and distance of 100 nautical miles"; and furthermore, waters adjacent to those already described, which overlie "submarine areas Her Majesty in Right of Canada has the right to dispose of or exploit", i.e. waters over the Canadian continental shelf. Now this definition covers a very large area of the high seas. And though not actually claiming these waters as territorial, the practical effect is not readily distinguishable from the South American claims of 200 miles of territorial sea which have been the subject of general protest.

The substantive part of the Act provides:

(i) penalties and prohibitions on the deposit of "waste" in Arctic waters, on islands or mainland. Waste is defined broadly as covering any substance "that, if added to any waters, would degrade or alter or form part of a process of degradation or alteration of the quality of those waters to an extent that is detrimental to their use by man or by any animal, fish or plant that is useful to man . . ."

(ii) civil liability resulting from the deposit of waste by persons engaged in a comprehensive list of activities in the area, including mining and navigation. The liability is absolute and does not depend on fault or negligence. But the Governor-General may make regulations establishing an upper limit in certain cases.

(iii) the Governor-General may require evidence of financial responsibility from persons proposing either to mine or to navigate in the area.

(iv) the Governor in Council may make regulations about the design, etc., of ships to be allowed into certain zones, about safety equipment, pilotage, ice-breaker escort, etc.

(v) the destruction or removal of ships in distress depositing waste or likely to do so.

(vi) seizure of any vessel suspected on reasonable grounds of suspected contravention of the Act. The seizure may be in Arctic waters or any other Canadian national or territorial waters.

This then is legislation at last which really faces the problem of oil pollution and is probably adequate to the job. But it is limited to Arctic waters, and this is its offered justification in law. It makes very considerable inroads upon the freedom of the high seas, even for the traditional use of navigation. It forms in many ways an awkward precedent. And it has already been the subject of vigorous protest from the Government of the United States.

What conclusions should be drawn from this rather sketchy survey?

On the international law side an impressive apparatus of paper has been produced within quite a short space of time. Whether it is sufficient and effective is another matter. The simple experiment of attempting to sit on any English beach is likely to dispel any complacency. And the likelihood and possible scale of new pollution is increasing by several factors almost daily.

The Canadian rather drastic step of simply extending comprehensive domestic legislation with severe penalty clauses to the high seas may probably be more effective. But this precedent could hardly be followed for temperate waters because other countries would not stand for it. Indeed they don't seem to like the Canadian move into Arctic waters much either; but I think they will have to accept that.

I have one suggestion to make in regard to legal techniques. The need is to bring international and domestic law closer together and particularly to strengthen thereby the sanction of international regulation. I have always thought that the ingenuity and perseverance of United States courts in finding arguments for applying their antitrust laws to foreigners for activities conducted outside the United States, was worthy of a better cause. The objection to the extra-territorial enforcement of antitrust law is that much of it is peculiar in purpose and content to the United States. But there is great attraction to me in the possibility of wide domestic court sanctions applied to domestic law which embodies fairly agreed and accepted international law. By some such technique it might be possible to achieve for the high seas something not unlike the effect of the Canadian legislation for Arctic waters. This is not entirely a new idea, of course. At a time when the common law countries hotly rejected the idea that any state might declare a "contiguous zone" of jurisdiction extending on to the high seas, the common law courts had no difficulty in actually extending their jurisdiction on to the high seas, sometimes beyond any possible contiguous zone, by the ancient legal device of "constructive presence", *viz.* being deemed to have been in a place where you were not. And, after all, tankers must eventually enter the ports of their customers.

These techniques have been used powerfully by United States courts to sanction not only antitrust, but labour laws, regulations governing securities, and much else. If they can be used to enforce national policies, could they not be used to enforce international policies?

Another possible step in busy waters, suggested in the UK Government's Note on "Lessons arising from the Incident of the *Torrey Canyon*", is to require tankers of a certain size to keep to defined sea-lanes when they are near a coast. After all, no one would contemplate allowing the captain of an approaching aircraft to decide in his own discretion his route, altitude and speed. Is there not a case for looking again at the extraordinary freedom allowed to the master of a ship? As far as I know the only experiment so far has been to establish a voluntary recommended rule of the road in the narrows of the English Channel.

Of course, however it is done, the cost, which must be considerable, will have to go on the cost of oil. So we must be willing to pay as well as to talk.

Discussion

PROFESSOR P. F. WAREING, CHAIRMAN

Obviously a number of extremely interesting questions have been raised by the speakers, and I feel it would be appropriate if the Chairman were an international lawyer rather than a biologist, but in the circumstances I will have to do my best. Of course, I do not want to control the discussion too closely, but, as our Chairman this morning said, one wants to get some sort of order into the discussion.

I have, therefore, picked out a few highlights of what particularly struck me in the three papers and, by way of a start, I would select from Professor Clark's paper the point that he made towards the end that the effects of pollution may be rather insidious and not apparent for a long time, and that by the time they are evident, the damage may be irreparable. The second point that I noted in his talk was that he suggested that legislation should keep ahead of evidence. Then Mr Holdsworth's text seemed to be that industry provides the services that the community requires and that, in this connection, if the community desires pollution control it must pay for it. But, on the other hand, it may decide, as in the case of the Bavarian refineries, that the saving in the cost of oil is worth a deterioration in the environment. Going on to the paper by Professor Jennings, read by Miss White, this, of course, raised a considerable number of points, but in particular I think they can be put under the headings of: the need for rules to prevent pollution happening; rules to contain and mitigate disasters; and rules to cover damages. Now, I think that if one started off by getting involved in some of the general questions raised by Professor Clark and Mr Holdsworth we might find ourselves ranging rather far afield, so perhaps we had better open the discussion on some of these major points of principle first, but I hope that we will not spend an undue proportion of the time on these aspects, but try and get down to some of the concrete suggestions that were raised by Professor Jennings' paper. So, may I suggest that somebody opens the discussion.

PROFESSOR E. EISNER

I am a physicist. I would like to say something which I think is relevant to every aspect of pollution. It has been touched on at various points, but I think we might make it more quantitative and follow a rule. I am going to quarrel, if I may, with, for instance, Dr Marley's way of presenting this information. Making a statement about the rate at which a pollutant is produced or distributed biases the problem. Any pollutant's effect is going to be a function of its specific toxicity – which will, of course, depend on the system one is looking at – it will be greater the greater its concentration and that, again, will depend on the size of the system one is talking about. It will obviously be greater if the rate of emission is greater, but the equilibrium concentration it will reach will be determined by the rate of emission times the decay time, or the time constant in the radioactive case. Thus:

$$\text{Pollution} = (\text{Specific Toxicity}) \times (\text{Concentration})$$
$$= (\text{Specific Toxicity}) \times [(\text{Rate of Emission}) \times (\text{Decay Time})]$$

I can conceive of no case where we should be talking about the rate of emission in assessing a pollutant: it should always be the product of rate of emission and decay time. The importance of this lies in Professor Clark's point that where decay time is very long we have to be awfully careful how we interpret rate of emission and toxicity. Since these quantities can vary by huge factors, decay time can be a governing factor. It may indeed be that, technically speaking, and perhaps even legally, the easiest way of tackling this problem is by concentrating on the decay time. In other words, saying that if something disappears from the system soon enough we need to worry much less about it and concentrating control and legislation on very stable pollutants. From this point of view it may be that the most important pollutant is not radioactive material or oil but polythene. It lasts 10^{10} years as far as we know!

MR A. R. TINDALL

Is it not even more complicated than this though?

PROFESSOR E. EISNER

Oh, much more. This is simplifying things.

MR A. R. TINDALL

The decay time which is relevant for a biological system can depend on the generation cycle of the organisms you are looking at, and what you have really got to be concerned about, I think, is the effect on the gene pool during one generation, or during a given number of generations.

PROFESSOR E. EISNER

By decay time I do not just mean physical decay time; it can be any kind of decay time, e.g. time taken to collect and destroy.

MR A. PRESTON

In terms of considering this concept of decay time, the ICRP recommendations for an acceptable level of exposure and an acceptable intake of radionuclides, employs an estimate of toxicity. This concept of toxicity takes into account not only the energy of the radiation from the radionuclide of interest, but also its decay time. The effect of decay is therefore built into the ICRP level.

PROFESSOR E. EISNER
But this is not so for many other pollutants and this is what I meant to convey.

PROFESSOR R. B. CLARK
From a biological point of view, an important factor to be taken into account is the persistence of a contaminant, that is, the length of time it will remain in the environment before it decays into something innocuous. To this must be added the possibility of biological accumulation. Persistent materials accumulating in biological systems are liable to become more and more concentrated as they pass up the food chain and can easily reach dangerous concentrations in some organisms because of this. Persistent, accumulating materials are the most dangerous toxicants that can be added to any environment. We have seen this already in the case of organo-chlorine pesticides and the same may be true of a number of other substances. There may not be a great many of them, but heavy metals must certainly be included in this category.

Because these materials are not readily degraded and therefore last a long time, and particularly if they are accumulated in a biological system where they may reach very high concentrations, their concentration and the dilution factor at the point where they are discharged into the environment bears very little relation to the situation that develops within the biological system.

PROFESSOR J. F. GARNER
As a lawyer could I just ask our scientific friends something about the methods of dealing with oil pollution? I understood, as a pure layman, that some of the things that were done to the beaches as a result of the *Torrey Canyon* in fact had a very much worse effect than the oil itself. I do not know whether this is true.

DR C. G. DOBBS
While we are on this point of the uncertain effects, I should like to link this with the criteria we use in determining "safe" or "control" levels. Lawyers and politicians depend on scientific advice, and scientific advice comes mainly from the specialists in the industry, technology, or whatever it may be which is causing the pollution. Their evidence is very often, though not always, accurate; they give a great impression of definite knowledge up to the point of the emission, and they have a great deal of information. After that the information is quite speculative, very often necessarily so. Once your pollutant has gone out into the general environment you are in a field of speculation, however much work you do. You then have the weighing of benefits, supported by a lot of accurate information, against vague fears which, nevertheless, quite often turn out to be justified, based on inaccurate and uncertain information and speculation from it. We have, for instance, referring to this morning's discussion, an extrapolation from lethal damage due to high levels of radiation to the effects of low levels of radiation which may well cause no lethal damage; and what damage they may do, if any, is lost in the pool of general damage. Now because all biological statistics have

a considerable range of variation there is a pool there in which you cannot make any statistical judgments – a pool of error due to normal variation. This does not mean that there is no effect, but merely that we have no means of isolating it. On top of that there are synergistic effects of which we know practically nothing. If you treat each pollutant separately you may think you have what we call a safety threshold. But we know very little at all about the synergistic or additive effects of the collection of different pollutants affecting any organism. So when we hear about safety factors of 10 and so on, which sound very impressive,they may be, in fact, quite inadequate.

Dr W. R. P. Bourne

I think, if I could put the ornithological point of view, we live in a different universe; we started having *our* problems in the eighteenth century; we have been at war since 1920, before I was born. The Adam Report to the Royal Society in 1936 was comparatively late literature to us, and I think we have done quite a lot to create the climate in which this sort of discussion takes place. Yet it seems to be a dialogue taking place between physicists and lawyers without much consideration for the politics involved, because much of the basis on which the battle is fought is that of influencing public opinion to pay some attention to the problem, and, in particular, to bring people into the frame of mind in which they will accept legislation. The public at large must be prepared to force it on commercial interests, who are not particularly malignant, but who have to consider their profits and who are liable to incur very great expenses.

From the bird point of view I think the reason why we are in the front of the battle is because birds and water-boatmen are the only things that habitually sit on top of the water, while oil floats. This is, of course, why oil has been the first major source of pollution to attract attention before practically anything else, before radioactivity, before chlorinated hydrocarbons – it is the most obvious of the lot. Bird bodies float, and they last much longer than any other form of body, so that they come ashore, and when they are noticed people get upset about them. So our part of the battle has not been so much scientific, most of the time, as making use of our completely disproportionate impact on public opinion. Really, unless we make sure that full publicity is given to our particularly obvious problem, nobody is ever going to put appropriate corrective legislation through.

In a meeting like this, part of the problem is getting the message to a sufficiently large number of people so that they will give support. Presumably, there are legislators here, so this will get through in important places, but it is the public at large which is most important and, in particular, seeing that they are correctly informed. Another of the problems is that they are exposed to some very funny ideas. After the *Torrey Canyon* the news media got the notion that oiled birds were news. There are a wide variety of pollution incidents and it is frequently the wrong ones that get attention. The outstanding example this year was a very minor slick that was seen by a passing journalist from a helicopter and filmed. As far as one knows it caused no damage, but perhaps the outcry compensated for a greater disaster that received no publicity at all.

To come back to the problem as we see it, it is certainly, as Professor Clark said, the chronic pollution which is the greater problem. The birds which live

86

out at sea are those which, naturally, have a very low reproductive rate and chronic pollution prevents their recovery whenever there is major damage to the environment. In the case of natural disasters, they normally recover over the course of a year or two. If they have lost their small margin for recovery due to chronic pollution, whether it is a natural or unnatural disaster, it is one more step down the ladder.

Looking at the manner in which this problem has to be dealt with, it is very much a question of whether we can influence the oil companies. Obviously, legal sanctions are necessary, but one cannot have the whole of the oil industry in continual litigation. It has become ever clearer that a very high proportion of the casual incidents which occur are due to somebody's plain carelessness, and it really is up to the oil industry to try and control the few black sheep among their personnel who are responsible. In the daytime, the Board of Trade can catch a certain proportion of the culprits; at night it is quite impossible. An improvement in the situation really depends on sanctions imposed by the industry. It is largely a question of getting at the black sheep, the minor companies and the careless captains, and of finding some way of persuading them not to behave in such a way as to give them all, collectively, a bad name. It is a political problem as much as a legal one, although one needs legal teeth to deal with people who are caught.

PROFESSOR E. EISNER
I was on holiday on the coast of North Carolina some years ago and one morning we came out and the sand was polluted with oil, though it had not been so before. So I rang up the local coastguard station and they sent a man along in about an hour, who sampled it and said that this might provide a basis for a prosecution if the source of the oil could be identified – they know, of course, what ships there were in the neighbourhood that night. He went on to say that if people reported oil arriving on beaches within hours, one could prosecute most ships under existing national laws – in this case United States laws, but I expect the same applies in Britain – as soon as they put into port in that country. Is that, in fact, true?

DR W. R. P. BOURNE
I think the Board of Trade has various powers, but it becomes progressively more difficult to catch people once they get away from port. I think last year they brought something like 60 or 80 cases, and in all but about two cases the offence took place in port. The moment it gets out to sea the ship disappears rapidly and unless it is a case like the *Andes* where an airliner got the name of the ship, it is almost impossible to prove legally. In particular, an oil slick will drift a long way quite fast; if the wind changes its course is very complicated, and it becomes extremely difficult to trace its origin, especially to prove it in a court of law. One of the things that strikes me as being extremely unfair is when people criticise the Board of Trade, who are doing their best to remove slicks, often quite small ones, that have been drifting for days from somewhere like Denmark and that are not detected until dead birds come ashore. The slick drifts far faster than the birds, so that by the time that the birds come ashore, and the complaints follow, the slick has either been on the shore for a week or has drifted away to

Norway. It becomes extremely unfair to blame the authorities in such situations. I think, myself, that it is very much a question of internal hygiene in the oil industry. This is partly a question of the goodwill of the industry itself and sheer concern for their own good name.

Professor P. F. Wareing, Chairman
There were some figures given by Mr Holdsworth, namely that it is thought that if the new IMCO regulations could be enforced (and that, of course, is the big "if") pollution could be reduced to as little as 15–20,000 tons per annum on a world basis. However, we are now getting involved in the mechanics of how control can be effected, which is, I should have thought, an expert's job, although we might well discuss whether it really is feasible to enforce these rules.

Professor R. B. Clark
There is one point about the proposed legislation arising from some of Dr Bourne's remarks that I should like to raise. The proposals appear to be directed largely towards the oil tanker industry and it seems to be assumed that if the oil transport industry can be made to behave as carefully as possible, the problem of oil pollution will have been solved for practical purposes.

I would like to ask what proportion of coastal oil pollution in heavily trafficked areas like the English Channel and North Sea is caused by tankers and what proportion by other shipping? I have the impression that while the latter represents trivial pollution in individual cases, in aggregate it may be considerable. If this is so, the new proposed legislation seems unlikely to be any more effective in controlling this source of pollution than the existing legislation is.

Mr M. Holdsworth
It is very difficult to answer this very specifically and one would need much more statistical evidence than we now have. Somebody mentioned the process of taking samples, analysing them and then tracing the sources, but I am afraid that, of the samples that have been analysed, a very large proportion of them are traceable back to crude oil, though certainly there are some samples of fuel oils which could apply to any ship. A very large proportion of the tarry globules arriving on our beaches appear to be crude oil residues, so therefore it must be from tankers. It is also more likely to be from tankers rather than from other ships because, short of accidents, no cargo ship using bunkers wants to put oil overboard. It may, accidentally, by the turn of a valve, do so, but there is no operational reason for such action, whereas there is an operational reason for putting out some oil from tankers during the necessary cleaning operations. What the proportion is I do not know, but I would say that the majority of those responsible are tankers.

Mr S. L. D. Young
Mr Chairman, it may help the discussion if, at this stage, I make one or two small explanations with respect to the Convention and, in particular, to some points which were raised in the introductory speeches. First of all there was the point about the 20,000 ton ships, that is new ships, which, I think Professor Clark said, were required to retain all their oil aboard. I would like to explain that this is not quite the case. The Convention lays down a prohibition on the discharge

of oil or oily mixture. Now, this term "oily mixture" is defined in the Convention as being "one containing 100 parts per million of oil, or more", which does not provide for total prohibition in the absolute sense. This also applies to the prohibited zones which are now in force in that the prohibition only relates to mixtures containing 100 parts per million of oil or more. Under the amendments which were approved by the Assembly last year this definition has been changed to "a mixture with any oil content" so that when a prohibition is applied it will definitely mean a total prohibition of oil.

Another point with regard to such ships is the proviso in Article III (c) that "if in the opinion of the master special circumstances make it neither reasonable nor practicable to retain the oil or oily mixture on board, it may be discharged outside the prohibited zones". It seems possible, perhaps even likely, that a good many masters, at some time or another, have decided that it is neither reasonable nor practicable to retain a quantity of oil aboard, and have discharged it, thus giving rise to a considerable degree of persistent pollution. Now the new amendments are designed to restrict the rate of discharge to 60 litres of oil per mile travelled, and that means a uniform distribution. It works out, if my arithmetic is correct, to something like 10 c.c. of oil per foot travelled. The amendments also restrict the total quantity which the tanker may discharge in any one ballast voyage. This is the first time that such a limitation has been applied on the total quantity which will be one-fifteen-thousandth part of the total cargo-carrying capacity of the vessel. Thus a 150,000 ton tanker, for example, will be allowed to discharge something like 10 tons of oil, which is a very small percentage of what such ships have been able to discharge up to the present time. The amendments, when they come into force, should therefore have a significant effect first of all by preventing the dumping of large quantities of oil and secondly by distributing the oil that is discharged over a long distance. Furthermore ships, even within the prohibited zones, can at present discharge a mixture of less than 100 parts per million and can be stationary while doing so, thus causing quite a large slick of oil. This will no longer be permitted under the new amendments.

There was one other point where reference was made to the development of a compensation fund for pollution damage. A working group in IMCO is at present busily engaged on this. I think something like nine major questions have been raised on the application and implementation of the fund, such as, whether or not compensation should be paid out for pollution from natural causes, or in cases where the victim has contributed to the pollution through some negligent act, etc. These and similar questions are being studied in detail with a view to further action. Thank you, Mr Chairman. I shall be happy to give further information should any other points be raised.

PROFESSOR P. F. WAREING, CHAIRMAN
Actually it was Dr Miller, I think, who said that he could speak on the Canadian proposals. This might be an appropriate occasion for him to come in.

DR D. M. MILLER (Canadian High Commission)
I am not really prepared to speak on them, but I did listen rather attentively to what Professor Jennings said in his paper about the Canadian proposals and I think they are of interest to our discussion because it is a perfect example of the

sort of unilateral action that a country takes if, in its political judgment, it believes that the international climate is not moving the way that it thinks will protect its national interests. I would like to correct one or two impressions that arise from what Professor Jennings said; in the first place, the statute that he was referring to is not yet law. It has gone through our Lower House but not through our Upper House, and our Parliament has risen for the summer recess. It is therefore not likely that the legislation will become law for some months yet. Nevertheless there is a very strong determination on both sides of the House to put this Bill through, and this will be reflected in the Upper House as it has been in the Lower. The rest of the description of the Bill itself was quite correct; he described it as being drastic, and I think it is drastic, but it has to meet what is deemed to be a drastic situation. I do not like to take up your time, but if you think that there is some value in getting the atmosphere of why Canada, a country which is not really known for acting unilaterally in the field of international law, a country which has, moreover, been very active in the Law of the Sea Conferences, and has put forward some compromise proposals on territorial waters which were almost adopted, I would be quite pleased to read what my Foreign Minister said when he introduced this Bill in the House, because it does give the atmosphere of the sense of frustration.[1] It covers international developments as foreseen from Ottawa, and explains some of the reasons why Canadians believe that the Arctic needs protection, which the Bill is designed to give, primarily because of the Alaskan oil discovery and the interest which the oil companies have demonstrated through the, now, two journeys through the North West Passage by the tanker *Manhattan.* I hesitate to read it because it would take five or ten minutes, and I do not want to take up so much time.

PROFESSOR P. F. WAREING, CHAIRMAN
Would anybody like to comment on the suggestion made by Professor Jennings about extending this principle to other areas, that is, extending sovereign rights over the high seas?

MR P. ARMSTRONG
I expect some people know that President Nixon, on, I think, 23 May, made the statement about the Continental Shelf, in which he envisaged what is called the "Intermediate Zone", going from 200 metres sea-bed depth to the rest of the continental margin – he does not state what that margin is, so it is open-ended, again, at the other end – but that whole area would be administered in his proposal by the coastal state, subject to payments and rules laid down for the area of sea-bed beyond 200 metres depth. This, if it is going to be at all like the Continental Shelf regime, will of course have repercussions on the superjacent waters, whether it is meant to do so or not. So you have a proposal which might take over as much as 20 per cent of the whole ocean floor administered by the coastal state. This is, in a way, a comment on Professor Jennings' point about the Canadian proposal; the march is going on or, as some of you might say, "the grab" by nation states continues.

I would like to ask if I understood rightly Professor Jennings' paper, so well delivered by Miss White, why, in fact, these conventions – the mass of paper

[1] The speech and the Bill referred to in Dr Miller's remarks are reproduced in Appendix II.

which Professor Jennings described – only extend to coastal areas? Is not the whole point about the sea and about these slicks that they move around extremely quickly, and if you confine a convention to the coastal area then you are not really dealing with the problem. So it does throw some doubt on the whole exercise; secondly, if you want it to have an effect, then ought you not to have some better enforcement mechanism? I think, if I understood it aright, states should have inspectors of craft or vessels which may be dropping oil at random. I would like to ask whether the Ruritanian inspectors only inspect Ruritanian vessels or whether they have the right to inspect the ships of other nations? How is this inspectorate really worked, and is it likely to be efficient? A third point, and it is only a suggestion – a bit "way out", I suppose – but if the oil companies really do not like the impression which people get from all this oil seeping around, have they ever considered a sort of oil company Securicor; that they set up their own system of inspection? It is quite clear that international law "which the evil ignore, and the righteous refuse to enforce" will never get anywhere without enforcement and that, therefore, somebody else is going to have to do it. I would like to ask Mr Holdsworth of Shell whether they have ever thought of having their own enforcement system.

Dr G. White
Mr Armstrong talks about the Nixon idea of an Intermediate Zone, but I do not want to go into too much detail on that because I do not think it is germane to what we are thinking about. This word "administration" of the waters above the Intermediate Zone can be, I think, very misleading. The whole core principle of the Continental Shelf Convention as it now stands is that this gives the coastal state exclusive rights to explore and exploit the resources of the sea-bed and the sub-soil; and the status of the waters above the shelf, in so far as they are not territorial waters, so long as you are outside that limit (whatever it is), remain high seas. There are, of course, inroads; you do have your drilling rigs and installations with the zone around them, which obviously affects one's freedom of navigation and freedom of fishing, and so on – you have not got absolute freedom of the high seas; you never have had it. But I think it would be wrong to give the impression that either the existing Convention or any ideas or proposals that are being floated by the United States or other countries at the moment are intended to cut off and turn into territorial waters areas of the high sea. I think President Nixon was talking about the sea-bed, the sub-soil and the rights and titles to minerals and so on from those areas.

The other points Mr Armstrong raised. Firstly, he talked about the Convention on Intervention, I think, when he said, "Why is this limited to the territory of states and why does it not cover the high seas?" I am not clear about the answer to this; I have not had a chance to read through all the discussions that led to the Convention – perhaps Mr Young from IMCO can help here. I have a feeling that this was drawn up in this way first of all because it was what states were prepared to accept and, secondly, in the light of the 1969 amendments to the 1954 Convention on Pollution, which has already been described to some extent. If that comes into force, as Mr Young has explained, this will very greatly reduce the amount of stuff that can legally be dumped on the high seas. So it may be that if you put the two together it will be quite effective. In other words you are not merely protecting the coastal area and territory of the state because you will have to take into account the 1969 amendments to the 1954 Convention.

The third point you raised was about the inspectorate, and on that, I am afraid, I am not entirely sure I have got the text – I do not know whether anybody else has, perhaps Dr Bowett can say something about this.

DR D. W. BOWETT
I think officials of the flag states carry out inspections of their vessels in their own ports, of course. They can carry out inspections of their own vessels in a foreign port providing the foreign state agrees to their inspection. Is that right? Otherwise they have not a right to be there, and you rely upon the efficiency of the local inspectorate wherever they may be.

May I ask one question? Does this problem now arise from the vessels that do not operate the load-on-top system? Is that so? Why is not the most effective way of dealing with the problem, for the technologically advanced states – that is, the customers of these oil carriers – to ban their ports to such vessels?

MR M. HOLDSWORTH
On the question of the banning of ships which do not operate the load-on-top system, the difficulty is that until the new amendments to the International Convention are ratified, non-compliance with the load-on-top system is perfectly legal provided that the oily residues from the tank washing process are dumped in the non-prohibited zones (generally more than 100 miles from land). If and when the new amendments become law no ship will be able to carry black oil for any shipper/receiver combination which does not agree to accept the tankers washing residues ashore at either the loading or receiving ports. To do so without this agreement will put the tanker immediately into contravention of the Convention.

Dr Bowett's first question on inspection then becomes most relevant. Tankers dumping their residues in mid-ocean at night cannot be detected in the act but, if they arrive at their loading port with clean tanks but without appropriate residues on board, they must be in default of the law. Inspection of tankers arriving at loading ports and repair ports together with enforcement of the law by prosecution, when appropriate, is essential if the new amendments are to be effective. Governments were, through IMCO, presently investigating means for inspection and enforcement and it seems likely that the most practical method will be the creation of a small inspectorate which will examine tankers at random as they arrive in ports around the world, and then report offending ships to their flag governments who will take suitable punitive action.

PROFESSOR P. F. WAREING, CHAIRMAN
Presumably as tankers get bigger, and therefore fewer, the problem will become easier too.

MR M. HOLDSWORTH
Yes indeed.

Mr Harry Walters (Soil Association)
Professor Clark observed that disposal of oil into the sea does not necessarily mean its effective dispersal. The discussion today has indicated that this would be true in temperate and arctic zones. I hope this has not left the impression that oil may be dumped effectively in tropical waters. Three years ago I was travelling by sea from Colombo to Aden; the weather was abnormal, very hot and a continuous flat calm. As the ship came up the Indian Ocean, although several hundred miles south of Iran, I noticed that many square miles of sea was covered with a pink growth. The captain said that this was usually considered to be due to the dumping of oil by tankers but it could be observed only under flat calm conditions. We were far south of the Somaliland Drift and there was no sign of an oil slick. Apparently minuscule dispersal had provided a surface medium for the vast and unusual growth of microflora seen. It appeared that this was a demonstration of ecological unbalance on the high seas presumably connected with the disposal, but non-effective dispersal, of oil.

The oil companies have spent a great deal of time, money and expertise on films which draw public attention to many other forms of pollution involving fresh water and the atmosphere. When the measures described by Mr Holdsworth and Mr Young become fully effective I hope we shall see a film on the eradication of oil pollution at sea. When this is done examples such as the one I have quoted would provide visual evidence of the immense size of the problem.

Dr J. Carroz
This morning several participants deplored the fact that two nuclear powers were not a party to the 1963 Treaty Banning Nuclear Weapon Tests in the Atmosphere, in Outer Space and Under Water. I presume that the same problem exists with respect to the 1954 International Convention for the Prevention of Pollution of the Sea by Oil, perhaps even on a wider scale on account of the flags of convenience. I wonder whether we can be provided with figures on the exact proportion of the world tonnage that is covered by the 1954 Convention as amended.

My second question is more technical. I understand that offenders work usually at night, that is, ships dump their oil under the cover of darkness. I saw a report recently, I think in the *New York Times*, to the effect that a radar-like device had been invented and that it was being used off the coast of the United States by the coastal guard aircraft. The instrument, which is called a "microwave radio-meter", enables authorities to detect oil being dumped at night. I wonder whether this method is also used in the United Kingdom.

Mr M. Holdsworth
The present-day tanker is covered by the existing Convention. I think there are forty-two governments signatories to the Convention as it stands presently. These forty-two governments must cover, within the monitorised flag holdings, at least 90 per cent of existing tanker tonnage. It is to be hoped that the ratification of the new amendments will be at least as extensive and that, as a result, virtually 100 per cent of internationally trading tankers will be covered.

Regarding the detection and assessment of the seriousness of an oil slick on the sea from the air by means of micro-waves, this method is in its infancy and my understanding is that, at the moment, the method requires a feed of data from a number of sophisticated tools (radar, infra-red, ultra-violet, etc.) before a balanced assessment can be made.

PROFESSOR P. F. WAREING, CHAIRMAN
If a tanker is owned by a company which is anxious to abide by the rules, has the skipper any incentive to dump his oil and contravene the intentions of his company? Is there any reason why he should want to get rid of it in an illegal manner?

MR M. HOLDSWORTH
None whatsoever.

PROFESSOR P. F. WAREING, CHAIRMAN
So one cannot really pin the responsibility on the skipper as an individual who is trying to save a lot of bother or a lot of time or money or whatever?

MR M. HOLDSWORTH
Not at all. His problem is that if he is hired to deliver a cargo to a refinery and the refinery says it does not want any slops then he has no alternative but to dump his slops. Of course, at the moment, he does it because it is perfectly legal in the non-prohibited areas.

DR J. W. HOPTON
May I ask a particular question? What are these tankers being inspected for? And, secondly, this gentleman mentioned the rate of release of the oil – 5 ml every yard – is this a figure plucked out of the air? What does it mean in terms of release? Where does this recommendation come from?

MR M. HOLDSWORTH
Basically, a tanker under the new amendments will be inspected on arrival at a loading or repair port in order to ascertain whether it has aboard consolidated residues commensurate with the tanks which had been cleaned en route. Fortunately by simple examination of the condition of the ship's cargo tanks, its Oil Record Book and its official ship's log it should not be difficult to discover the tanks which have been cleaned, the geographical location of the cleaning and the presence on the ship, or otherwise, of the consequent residues.

In the actual cleaning operation the new amendments permit an effluent of oil content of up to 60 litres per mile of ship travel. Compliance with this criterion is not so easy to ascertain, but a knowledgeable inspection of the Oil Record Book and ship's log would certainly reveal whether the ship has complied with the spirit of the rules. The figure of 60 litres per mile has not been plucked out of the air by IMCO in framing the new rules, but has been based upon extensive experiments over a number of years by Her Majesty's Government. These experiments show that oil put out from a moving ship at such a rate will very quickly spread to a very thin film, which will itself disappear completely from the sea surface in two or three hours by process of natural dispersion, presumably into the waters beneath. Although such oil spreads and disperses quickly beyond means of simple detection, its final degradation by microbiological attack, chemical oxidation, etc., is likely to take somewhat longer.

Professor P. F. Wareing, Chairman

May I ask whether it would be true to say that you feel that if these new proposals are accepted (a) they will be very effective and (b) they can be enforced?

Mr M. Holdsworth

The potential effectiveness of these new amendments lies in the fact that they will be enforced. The present Convention is ineffective because it cannot be enforced. At present a ship can be photographed putting out a large amount of oil, but the slick from the air will not necessarily look significantly different from one produced by an effluent containing less than 100 ppm of oil which is presently allowable anywhere outside territorial limits. In the new amendments no oil from a tanker would be permissible within 50 miles from shore, so that any slick detected behind a tanker in such waters would indicate a contravention of the rules. In addition, as mentioned earlier, the requirement in the new rules for a tanker always to arrive at its loading port with its washing residues still on board provides an enormous step forward in the possibilities for inspection and enforcement.

Miss M. M. Sibthorp

Can I just ask one small question? I read that Thor Heyerdahl was sailing across the Atlantic in his ship, and reported that for mile after mile in the mid-stream of the Atlantic he found smaller or larger lumps of solidified asphalt-type oil floating. Now what could have caused that and would it do any harm?

Mr M. Holdsworth

I have to admit again that it seems most likely that this type of pollution emanates from the presently legal disposal of tanker washing residues at sea. As I have said earlier, it is recognised that at the present time something like 600,000 tons of such oily material was being put into the sea. As to what harm the presence of such solid tarry lumps might do at sea, perhaps Professor Clark would comment. It seems to me that the solid lumps, having once formed, will not spread or disperse into the sea and therefore will do no harm to marine fauna or flora. The harm these dirty lumps do to our tempers having arrived eventually on some amenity beach is all too well known.

Lord Simon

I read the same report, and it seems to me that as Thor Heyerdahl was drifting across the ocean and he saw it day after day, this was due to the fact that he was drifting with it.

Dr W. R. P. Bourne

When Miss Sibthorp pinched that particular point, I was just going to use it. To go back to radar, I have done quite a lot of work with radar and with particular types you can see the pattern of the waves rather well. I would have thought that you frequently ought to be able to see an oil slick, but the problem would then be to confirm that the slick was oil and catch the ship that did it while it was still within range and you could prove it was the right ship. I would have thought you could do something then.

95

Then to turn to this business of the tropics. Of course, we are so busy legislating all the oil out of European waters that – I think one must give the credit to the oil companies – despite steadily increasing traffic the average background pollution has become no worse than before. The accidents are getting more and more horrible, but these are accidents. The net result of the improvement at home is liable to be that people do their washing-out somewhere else. After the Suez Canal was closed for two or three years there were a whole series of incidents off South Africa which have caused very serious concern for the status of the local penguin, and so on. I do not know if that is still going on, but the thing one immediately suspects is that following the outcry in the South the ships must now be washing out in the tropics where there are vast areas of sea and people are not sufficiently organised to do very much about it. Personally, I have been rather concerned by what may be happening in remote parts of the tropics and I have been in touch with various people to see if they knew what was going on, particularly in the Persian Gulf. There are quite a lot of bird watchers in the Persian Gulf now and they see a certain amount of oil on the shores at places like Kuwait. In point of fact, as far as birds go, one does not hear of much damage. I think this is partly because the sort of swimming birds that get into trouble are not very numerous in the tropics, though there are some in places like the Gulf. It may also be, with the much higher water temperature, that bacterial decomposition of the oil is very much more rapid and the evaporation of the volatile components in the oil, which include a lot of the most toxic ones, is also much more rapid, so that, in point of fact, the oil is degraded more rapidly in the tropics. Also, of course, there are very much larger areas of the tropical sea in which it can get lost. It seems to me that the worst disadvantage to getting rid of oil in the tropics, if it has to be washed out somewhere, is that the tankers apparently tend to blow up while washing out.

Dr D. W. Bowett
I thought there had been the suggestion that you might add a trace element to an oil cargo which would help you to identify the ship which has discharged the oil. Is that so?

Mr M. Holdsworth
I agree that technically a radioactive label could be imprinted in a tanker's cargo so that, if the cargo is spilt, the cargo and ship could be identified. However, with the enormous volume of crude oil trade, the technical and administrative problems of tagging each and every cargo in a fully discriminatory way would be quite immense. It seems to me that, if effort can be put into enforcing the new amendments to the Convention, it will provide a much more viable way of preventing pollution than by cargo labelling. In the case of pollution from a tanker accident, there is rarely any difficulty in identifying the polluting ship. Such accidents almost invariably require outside help and, as a result, the ship and its misdemeanour become public knowledge very quickly.

Dr J. A. G. Taylor
It is not a question of treating your ships like factories. There is no reason why you cannot treat the effluent discharged – the discharge level of 60 litres per mile

96

allowed is quite thick on a molecular scale if you spread it out. You have got quite efficient treatment procedures in your factories, why not in your ships?

MR M. HOLDSWORTH

You said that 60 litres per mile is quite a thickness but, in fact, it is rapidly transformed into molecular thickness. It is almost immediately giving irridescent rings which turn very quickly into an overall grey sheen which, in turn, within an hour or two disappears without further visible trace. Although it is possible in a refinery to reduce where necessary oil content of effluent to 10 ppm or less, the space required and the residence time involved is quite beyond that which would practically be made available on a trading tanker.

DR J. A. G. TAYLOR

Could you not treat with the carbon you can generate from the Shell pelletisation system using absorption?

MR M. HOLDSWORTH

Well, one could do many things, but you must remember that if ships are going to be reasonably economic they cannot have half their space carrying cargo and the other half carrying equipment.

DR E. WINDLE TAYLOR

The question really is, are you going to ask the public to pay another halfpenny a gallon to pay for this kind of process?

MR M. HOLDSWORTH

The real problem is that there is an enormous amount of water to be dealt with. When we are talking about the total oil content, as we have been doing – Mr Young was talking about the limit on the total amount of oil put out from the ship – we are counting in minute quantities, around about 30 parts per million, in the ballast water which must be discharged for operational reasons. Now a 210,000 tonner, which is now a typical sort of large tanker, may well carry something like 80–100 thousand tons of ballast. To be able to treat 100,000 tons of ballast water in the operational time available, even with advanced methods at the moment, is an enormous problem which, frankly, I do not think we are capable of solving.

DR J. A. G. TAYLOR

Over what period of time would a million gallons of water be discharged?

MR M. HOLDSWORTH

True, if you had a ship that could go on and on and trickle 100,000 tons of ballast over a long period, you could design some reasonably small equipment to deal with it, but that 100,000 tons of ballast, for operational reasons, simply has to go out at, probably, 8,000 tons an hour, otherwise the ship just cannot get clean and do its job before it gets to the loading port. Now, if you try to deal with 8,000 tons an hour of water, and reduce its oil content from 30 parts per million down to 10 parts per million, you will have a problem on your hands.

SIR FREDERICK WARNER

On the point of where this 60 litres per mile goes to and how long it takes, I would have thought that this particular factor is of relevance, and my own understanding – perhaps a microbiologist here could confirm this – is that Philpel at Chelsea had demonstrated that this amount was completely oxidised within 24 hours – in temperate waters.

Dr J. R. Goldsmith
With respect to the legislation being proposed and discussed in Canada, it seems there are certain ecological and climatological differences that affect the problems there, but are there not many other countries which are at least involved in or affected by similar problems, such as the countries along the northern tip of Europe? Possibly, even the problems associated with the passage of tankers around Tierra del Fuego may have some similar attributes. Would it not, therefore, be worth while to consider special regulations – perhaps they have been considered – for transport of oil and tankers in Arctic and Antarctic waters?

Professor P. F. Wareing, Chairman
Mr Miller was about to raise a point anyway. I do not know whether you want to comment on this point as well?

Dr D. M. Miller
Perhaps I can, but I would like to comment first on the results of the Brussels meeting in October 1969. The impression is given that what was done there is going to solve satisfactorily the problem of oil pollution on the high seas. It was the view of my government, and still is, that the results of the Brussels meeting were not all that satisfactory. You will recall that we heard this afternoon that there was a public law convention and a private law convention. As to the public law convention, it represents quite a step forward in the case of accidents; it does give the coastal states the right to intervene and indeed, if necessary, to sink the foreign vessel causing the pollution of the coast. But, strangely enough, the Convention did not give the right of a coastal state to set, in advance of such an accident, some safety standards that it deemed should apply to the coastal area around its territory. So that you have the right to sink, but you have not the right to prevent.

There is another aspect which needs to be borne in mind. In the private law convention which, I understand, deals with liability, the principle was still not accepted that the carrying of crude oil, or refined oil more particularly, in large quantities such as the world is witnessing now with the construction of these enormous tankers, is really an ultra-hazardous activity. That means, in legal terms, that there should be an absolute liability to recompense in full for the damage that occurs. In Brussels, as I understand it, a ceiling was placed on the liability to be applied; I believe we heard from one of the speakers that this ceiling is something like 10 million dollars. One only has to reflect on the *Torrey Canyon* disaster to try and put that amount into perspective, and then to imagine what would have happened had the *Torrey Canyon* been one of these larger tankers which are under construction today.

As to the point about the legislation in the Arctic, the Bill that is in the process of going through the Canadian Parliament establishes an Arctic Pollution Zone, it creates a Board to regulate for that zone, and it is the intention of the Government to consult, in advance of the drawing up of regulations, those countries which are known to be interested in the Arctic from a user's point of view. It will be a fairly extensive list of countries that will be consulted and I would expect that some time, perhaps later in the year, you will witness some sort of conference taking place, perhaps in Ottawa, of experts who will look at the problem of navigation through the Arctic and draw up navigational and safety standards to be applied in the Arctic by agreement.

I have nothing to say, really, about the Antarctic because I know nothing about it – indeed I know nothing much about the Arctic – but I would imagine that, conditions being similar, any spillage there would be as bad as in the North in terms of the damage that it would do to the ecology of the region. This is really the main thrust of the Canadian legislation, cynical comments about sovereignty notwithstanding, and nobody, I think, seriously challenges the Canadian jurisdiction in the North – you have only to go there to realise why – it is a matter of protecting this part of the world, in terms of what is there, its natural balance, the weather aspects and so on.

Professor P. F. Wareing, Chairman

You were pointing to what you regard as the weak points of the recent Convention in relation to accidents, whereas Mr Holdsworth was arguing that, in terms of normal unloading, it could effect a great improvement and he was inclined to say that the overall effects of accidents should not be exaggerated.

Mr M. Holdsworth

May I just qualify that? I think one has to say "accidents in temperate zones". I must say I fear there may be problems in the Arctic where we just do not know to what extent biodegradation can be a factor in the removal of spilt oil. An accident in Arctic waters is going to be something the consequences of which we do not know and which could be extremely serious.

Dr W. R. P. Bourne

Another subject I do not think we have ventilated much in the discussion is under-sea drilling. Union Oil, of *Torrey Canyon* infamy, pricked a bubble off the coast of California and caused a lot of trouble. Recently, there has been another lot of trouble in the Caribbean where somebody obeyed the law and put a valve in their drill-hole, waited until people's backs were turned, and then pulled the valve out again and oil was released. Last winter oil was struck in the North Sea and, shortly after, the wind blew from the east and we had a lot of oiled birds on the East Coast. We do not necessarily imply that there is a causal relation here, but one is worried, and one wonders whether enough attention has been given to the question of under-sea drilling in our home waters, in particular because people are drilling at greater and greater depths, and it is presumably going to be more and more difficult to control anything that goes wrong in wells in very deep water.

Professor P. F. Wareing, Chairman

These spillages do occur pretty regularly – for example, in Milford Haven, I believe – and until recently the fines imposed have been ludicrous. Is anything proposed to stiffen up the penalties?

MR M. HOLDSWORTH
There are moves afoot to increase the present rather innocuous level of fines. Recently, Kuwait instituted a procedure by which a writ is immediately put on a ship suspected of pollution and the ship can then sail only after deposit of a bond of over £6,000. If the suspicion is later proved the ship must pay the costs of clean-up and forfeit all or part of the bond, depending upon the seriousness of the pollution. Undoubtedly this procedure has increased the degree of care which is taken whilst in Kuwaiti waters.

I think that a Harbour Authority, such as Milford Haven, has the autonomous right to raise the level of fines without involving legislation.

LORD SIMON
If I may add one word to what Mr Holdsworth said, I think the fines are imposed by magistrates.

DR D. M. MILLER
In Canadian legislation, in terms of fines, it is, I believe, $5,000 a day for a person and $100,000 a day for a ship.

The David Davies
Memorial Institute of
International Studies

The Department of International
Politics of the University College
of Wales, Aberystwyth

CONFERENCE ON

LAW, SCIENCE AND POLITICS:
WATER POLLUTION AND ITS EFFECTS
CONSIDERED AS A WORLD PROBLEM

Sunday 12 July
MORNING SESSION

Chemical and Pesticide Pollution

Chairman DR D. W. BOWETT
Queens' College, Cambridge

Scientist MR M. OWENS
Stevenage Laboratory

Administrator MR N. J. NICOLSON
Thames Conservancy

Lawyer PROFESSOR J. F. GARNER
Nottingham University

Chemical and Pesticide Pollution

MORLAIS OWENS (Water Pollution Research Laboratory, Stevenage, Herts)

The prevention of pollution is concerned with controlling the quality of polluting effluents discharged to a river so that their effect is not detrimental to the river use. The quality of polluting effluents is controlled by applying consent conditions to the discharges. Powers of effluent quality control are statutory and are given under the Rivers (Prevention of Pollution) Acts of 1951 and 1961 and the Rivers (Prevention of Pollution) Scotland Acts of 1951 and 1965. The factors taken into account when granting and applying conditions of consent to discharge include consideration of the pollution upstream of the discharge, the volume, strength, type, and duration of discharge, the flow rate and volume of the receiving river, the retention time of the river and chemical and biological recovery rates, and the distance downstream from the discharge of fisheries and of water extraction points for potable supply, irrigation, or industrial use.

Water pollution has been defined as occurring when the water "is altered in composition or condition, directly or indirectly as a result of the activities of man, so that it is less suitable for any or all of the purposes for which it would be suitable in its natural state".[1] The pollutants entering rivers and lakes can be divided into a number of categories; these include: (i) oxygen demanding wastes, e.g. domestic sewage, farm wastes, and effluents from the food and drink industries, (ii) plant nutrients from domestic sewage, industry, and agriculture; (iii) toxic substances which include heavy metals, insecticides, herbicides, and other toxic organic materials; (iv) silts, sludges, coal and china clay washings, and sediment from land erosion; (v) oils; (vi) hot water from industry and power stations; and (vii) radioactive materials. All of these can cause serious problems in the freshwater environment and all can seriously affect the use or amenity value of the receiving water. Some of these have already been dealt with in other papers given to the Conference and the present paper will therefore be confined to the problems of oxygen demanding wastes, plant nutrients, toxic substances (particularly pesticides), and oils.

Oxygen-demanding Wastes

When organic material in effluents is discharged to a river or lake, one of its most important biological effects arises from its breakdown by the activity of micro-organisms. This activity requires oxygen and therefore the dissolved-oxygen content of the river or lake will be reduced. The sewage produced by an adult person gives rise to a daily oxygen demand of about $\frac{1}{4}$ lb (100 g); this represents the total amount of oxygen dissolved in over 2,000 gallons (40 m^3) of water. Fortunately, conventional sewage treatment reduces this demand by 90–95 per cent, but it is easy to see that if too great a load of organic material is discharged the water will become deoxygenated, leading to death of fish; furthermore, the resulting anaerobic breakdown of organic material will give rise to evil-smelling products, thus further impairing the amenity value of the receiving water.

The capacity of the river to cope with an organic loading depends upon an oxygen balance between the demands imposed by the oxidisable materials on the existing oxygen resources and the rate at which the water can take up oxygen from the atmosphere. The rate at which oxygen is added to the water as the result of the photosynthetic activity of plants, and the rates at which it is removed by the respiration of the same plants and by the oxidation of mud deposits, also have to be taken into consideration. Attempts have been made to describe mathematically each of these processes of oxygen addition and removal, and to create a mathematical model which would allow the prediction of the effect of an effluent of a given quality on the distribution of oxygen in a river. It would then be possible to suggest the standard to which a sewage works should be expected to perform so that its effluent should not have any deleterious effect on the distribution of oxygen in a river.

There is really no reason, except perhaps an economic one, why rivers or lakes need be overloaded with oxygen demanding wastes. The processes involved are well known; the technology required to alleviate the problem exists, and present legislation is probably adequate to ensure that suitable conditions are maintained if standards are enforced.

There are, however, some organic materials which are not readily bio-degradable and do not exert any demand on the oxygen resources of a stream or lake but might impair water quality in other ways. Because so little is known about the possible long-term effects of such materials on people they must be viewed with some concern should the water be used for public supply; sensitive methods of analysis have to be employed to determine the precise concentrations present, and consideration has to be given to the development of special methods of treatment to remove them from the water.

Plant Nutrients

The enrichment of water by plant nutrients has become so great in many fresh-waters that blooms of algae develop annually. These can give rise to objectionable tastes and odours, cause problems in water treatment by clogging filters, render lakes unsightly, cause deoxygenation, and also produce toxins. This is not a new problem for the "breaking of the meres" which foretold national disasters in the, Middle Ages were algal blooms; the Metropolitan Water Board has been coping with the problem for more than thirty years, and a report[2] published in 1947 cited thirty-seven lakes in Europe and the USA which during the last 100 years had become enriched with plant nutrients. Only fairly recently, however, has the extent of the problem and its consequences received national and international attention.

We have been lucky in the UK, in that our lakes are fairly remote from centres of population and therefore receive little direct discharge of untreated sewage. Increased concern here about the effects of enrichment stems from the increasing need to obtain water from more and more surface sources and from a desire to develop the amenity afforded by rivers, canals, lakes, and multipurpose reservoirs. Although enrichment of river water with plant nutrients does not generally create problems in flowing waters in the UK, it becomes of particular significance when such rivers discharge into lakes or impoundments or when water is pumped from them into storage reservoirs.

The qualitative needs of several algal species have been established, but it is clear that the requirements for any specific substance are considerably affected by the concentrations of other nutrients and by physical characteristics of the environment such as light intensity, temperature, and size and shape of the lake basin or reservoir. (Many specific nutrients have been studied including phosphorus, nitrogen, silicon, potassium, carbon dioxide, trace metals, vitamins, and other organic compounds.) Most attention has been paid to nitrogen and phosphorus in the belief that of all the major nutrients those most commonly limit growth. Nevertheless in some circumstances other nutrients may be more responsible for promoting blooms; for example, there is considerable controversy at present about the role of carbon dioxide.[3] As far as I am aware no firm proposals for the establishment of limits on the level of plant nutrients in effluents has ever been made, doubtless because in the present state of our knowledge and with the great variety of possible nutrients it would be almost impossible to justify such criteria scientifically. However, attempts are being made to define the concentrations of nitrogen and phosphorus below which the extent of algal blooms might be limited, and the results of these have been summarised in a recent publication of the OECD.[4] Plant nutrients are derived from many sources within a catchment area. These can be categorised as:

(i) *Diffuse sources* such as rainfall, run-off and drainage from agricultural and urban land, droppings from animals and birds, and leaf fall; and
(ii) *Point sources* such as sewage effluents and industrial wastes.

It is easy to estimate the magnitude of point sources and certainly they would be easier to treat and control when attempting to alleviate the problems caused by enrichment.

Sewage effluents contain high concentrations of nitrogen and phosphorus. The phosphorus is derived from excreta and in recent years this has been augmented by the phosphate derived from packaged detergents. It has been estimated[5] that in the UK between 30 to 50 per cent of phosphorus in effluents is derived from packaged detergents, in the USA 50 to 75 per cent,[6] and in Germany about 36 per cent.[7] These figures indicate that even the complete replacement of phosphates in detergents by some other material would not prevent the discharge of phosphorus in sewage effluents. Nevertheless, in view of public concern in Scandinavia, USA and particularly in Canada, where no phosphate-containing detergents will be allowed for sale after 1972, efforts are being made to find acceptable substitutes. The best known of these is nitrilotriacetate (NTA), which is being used on a fairly large scale in Sweden and is said to be capable of replacing up to 70 per cent of the phosphorus in detergents. The introduction of strong complexing agents such as NTA for large-scale domestic use must be approached with caution, for while phosphorus may be important in promoting algal growth, no directly adverse effects of this element on rivers and lakes or on the consumer of the water appear to have been reported. There is little experience of the effects of NTA, and while complete degradation might be achieved at efficient treatment works it seems likely that some will find its way into rivers from the less efficient, or during very cold weather, or from storm overflows. There is also the possibility that if NTA is added to sewages containing industrial effluents, then toxic metals, which are usually removed in the sludge, may be carried through the works as complex ions. In river waters NTA might alter equilibria involving trace elements and consideration should be given to its possible effects on metals used in the water distribution systems. It is highly desirable, therefore, that any proposed alternative to phosphate should not only be very readily degraded but that it should also be shown to have no adverse effects on the quality of river water, or potable water supplies.

Although considerable quantities of phosphate fertilisers are applied to agricultural land the phosphate is rapidly absorbed by most soils. Sales of phosphate fertilisers have remained fairly constant during the last ten years and it is therefore unlikely that the increased concentrations of phosphorus found in some rivers are derived from this source. However, when erosion of soils occurs considerable quantities of phosphorus may be transported by rivers.

Drainage from agricultural land in this country, the USA and Europe contributes considerable quantities of nitrogen to waters, particularly in the form of nitrate. In England and Wales proportions ranging from 5 to 99 per cent of the total nitrogen load carried by rivers have been estimated as being derived from land;[8] the exact proportion will depend upon the quantity of sewage effluent carried by the river. It has been estimated that a population of 3 to 4 persons per hectare is required before the contribution of nitrogen from sewage effluent equals that from land drainage.[8] Contributions of nutrients from catchments represent the quantities derived from rainfall, animal manures, applied fertilisers, and processes of mineralisation in the soil itself, and it is difficult to make direct estimates of the magnitude of the contribution from each of these sources. It has been widely suggested that the increasing concentrations of nitrogen in some river waters are the result of the great increase there has been in the use of artificial nitrogenous fertilisers. While the amounts used in the UK have certainly almost doubled in the last ten years there does not seem to be a direct correlation between fertilisers used and the nitrate concentrations found in river waters. Some rivers show significant increases, whereas a great number of those looked at do not, despite a considerable increase in the quantities of fertilisers used in their catchment areas. Furthermore, it has been reported that the quantities of nitrate derived from unfertilised agricultural land are as great as those from fertilised land.[9, 10] No rise in stream nitrate levels has occurred during a thirty year period in which there was a three to tenfold increase in the crop fertilisation of a 150,000 acre (60,000 ha) area drained by the upper Rio Grande River, according to a report by the US Agricultural Research Service.[11] This is some of the first evidence that nitrogenous fertilisers may not necessarily contribute significantly to the pollution of rivers. Clearly the nature of the soil and of the crops and the farming practice in the catchment will all have a considerable influence. In particular, there is little doubt that intensive farming systems tend to create conditions in the spring in which there is some

106

loss of nitrate nitrogen,[10] and their growth will undoubtedly pose considerable problems in the future.

It may be of some interest to compare the quantities of nitrogen and phosphorus reaching the land in the United Kingdom from various sources. The average rainfall is about 25.6×10^{10} m^3 per year, the nitrogen and phosphorus contents of which are 1·0 and 0·05 g/m^3 respectively, and thus about 25.6×10^4 tonnes of nitrogen and 1.3×10^4 tonnes of phosphorus are deposited annually in rainfall. About 80×10^4 and 40×10^4 tonnes of nitrogenous and phosphorus fertilisers respectively are applied to farmland annually. Animals excrete about 80×10^4 tonnes of nitrogen and 15×10^4 tonnes of phosphorus per year, much of which is applied to land and could ultimately reach inland waterways. These figures help to give some perspective to some of the statements made about fertilisers.

The quantities of nitrogen and phosphorus entering surface waters or finding their way into the seas surrounding the UK are also of some considerable interest. The average loss of nitrogen from land is about 10 kg/ha/year and that of phosphorus about 0·3 kg/ha/year. The area of the UK is about 244×10^5 ha and therefore about 24.4×10^4 and 0.7×10^4 tonnes respectively of nitrogen and phosphorus derived from land enter surface waters annually. These figures compare with an estimated 18.2×10^4 tonnes of nitrogen and 3.5×10^4 tonnes of phosphorus that are discharged in sewage effluents to rivers annually. Much of this nitrogen and phosphorus will find its way to the seas around our shores.

There has been considerable discussion about removal of nutrients,[12] particularly nitrogen and phosphorus, from effluents and waste waters, and methods have been developed which are fairly effective – and indeed are in operation in the USA and Sweden and are planned for Canada. However, before decisions can be made as to the feasibility of controlling the growth of algae by this means, information will be required on:

(a) which nutrients are likely to be critical for the growth of algae in the particular body of water under consideration,
(b) the amounts of the critical nutrients contributed by point and diffuse sources in the catchment area, and
(c) a demonstration that the removal of the critical nutrient will limit the growth of algae significantly.

It should be pointed out here that nitrogen may have to be removed from water for public health reasons rather than for the control of algae, for high concentrations of nitrate in drinking water can cause methaemoglobinaemia in infants.

Toxic Substances

Extremely low concentrations of toxic substances like heavy metals, cyanides and phenols which are often present in industrial effluents and sewages can kill fish and the animals they feed upon. The numbers and types of toxic substances which may be present in industrial wastes are very great and the degree to which these are eliminated in sewage treatment processes is largely determined by the extent to which they are removed by sedimentation or biological oxidation.

Some improvements can be expected in the future, such as that resulting from the disappearance of effluents containing gas-liquor from plants manufacturing "town gas" because of its replacement by "North Sea gas". The effects of different concentrations of individual poisons and the influence of the varying physico-chemical conditions of the environment upon the toxicity have been extensively studied by organisations in various parts of the world. Most of this work has involved tests with fish. A number of fish are exposed to various concentrations of the poisons for a given length of time and the median period of survival of the fish at a given concentration of poison is determined. Generally a threshold concentration is observed – this is the region where a small reduction in the concentration of the poison results in a very large increase in the survival time of the fish. This is used to decide the concentrations of the particular poison which can be accepted by the fish and to help determine the consent conditions for the discharges. It is expected that a fish-toxicity test will be used increasingly by river authorities in the future, especially when unidentified poisons are present. While this procedure tells whether adult fish will survive in the river, it does not say whether or not the conditions would be suitable to support a good fishery, for example it does not say which fish will mature and breed, nor whether the eggs will hatch, nor does it give information on adverse sub-lethal effects. Nevertheless, these short-term tests must still be accepted in order to screen the very large numbers of chemical compounds being produced and to obtain a measure of their possible effects under a variety of conditions, though at the same time suitable steps should be taken to minimise errors of extrapolation.

Pesticides may enter rivers directly through rain, aerial sprays, control of submerged and emergent aquatic weeds, surface run-off, and from sewage and industrial wastes. Results of several wide-scale and intensive surveys of pesticide distribution in river systems have now been published in both the UK[13, 14] and the USA.[15, 16] Surveys made in the UK in 1965–66 at some 450 stations indicated that the average concentration of organo-chlorine pesticides for all of the stations sampled was 180 ng/l (about 2 parts in 10,000 million), the range being 14 to 3,054 ng/l. As might have been expected, the polluted rivers tended to contain higher concentrations than clean rivers, 10 per cent of the polluted rivers containing more than 500 ng/l while only 10 per cent of the clean rivers had concentrations greater than 130 ng/l.

Despite the considerable publicity given to pesticide concentrations in river water in the USA, it seems from these surveys that concentrations in some British rivers exceeded those found in the USA. The highest levels found during the American survey were 4 ng/l γ BHC, 100 ng/l Dieldrin, 150 ng/l DDT, and 30 ng/l TDE, while in Britain concentrations in rivers in parts of industrial Yorkshire and Lancashire are 520 ng/l γ BHC, 540 ng/l Dieldrin, 1,330 ng/l DDT, and 490 ng/l TDE, and in the Rivers Stour and Severn (Warwickshire) 2,480 and 560 ng/l Dieldrin. These high levels were associated with industrial usage of insecticides for mothproofing. Insecticide residues from agricultural use were generally below 50 ng/l though some concentrations as high as 170 ng/l were found. While from this comparative viewpoint the concentrations in some British rivers may give rise to some alarm, recent experiments conducted by the Freshwater Fisheries Laboratory of the Ministry of Agriculture, Fisheries and

Food suggest that between 1,000–10,000 ng/l of either DDT or Dieldrin is only lethal to trout if this concentration is maintained for three months.[17] The toxicity of γ BHC is of similar magnitude. Such laboratory data do not indicate the total insecticide hazard to fish, since food organisms containing insecticides also constitute a potential danger; furthermore, it seems likely that sub-lethal concentrations of insecticides could contribute to the overall toxicity of the water, thus reducing the permissible concentrations of other poisons. American data also suggest that concentrations in bottom sediments might be at least 100 times those in the overlying water and little is known about their effects on organisms living within or feeding on such deposits.

The results of a recently completed survey of pesticide residues in samples of water, mud, and fish from six agricultural river systems in South-East England show that the organo-chlorine pesticides γ BHC, Dieldrin, and DDT and its derivatives occurred in all the rivers but concentrations were generally low – less than 60 ng/l. High peaks that were recorded did not appear to be related to any particular usage but rather to accidental discharge. The concentrations detected in water samples taken from some of the rivers previously sampled during the 1965–66 survey were generally less than those detected in the earlier survey. Organo-phosphorus pesticides were also found occasionally and persisted only for several weeks at a time; if the rivers studied were typical then these pesticides do not seem to present any persistent problem though there will always be accidents such as have been reported elsewhere with mecarbam in the River Lee, thiazinon in the Welland and Nene River Authority area, and phorate in Chichester Harbour.

Although DDT has been in use since the mid-1940s and considerable information is available describing its effects and those of other insecticides on aquatic invertebrates and fish, there are only limited and conflicting data concerning their effects on algae. It has been reported[18] that concentrations of DDT in excess of $10 \mu g/l$ inhibit the photosynthesis of marine algae while $0.3 \mu g/l$ has been shown[19] to inhibit the growth of and induce morphological changes in the green alga *Chlorella*. On the other hand, concentrations ranging from 1 μg to 100 mg/l have been reported[20-23] as having no effect on the growth of many species of algae including *Chlorella*. Some recent tests have shown[24] that $5 \mu g/l$ of dieldrin, aldrin, and endrin suppressed the uptake of oxygen by *Microcystis aeruginosa* whereas DDT had no effect at a concentration of 1 mg/l. Data from the same experiments showed that DDT, dieldrin, aldrin, and endrin at concentrations of 1 mg/l had no significant effects on the respiratory activity of three other species of algae. All of the data indicated that the algae could concentrate the pesticides many times and therefore the possibility exists that concentrated residues could be transferred along the food chain.

Since persistent organo-chlorine insecticides occur in unsprayed areas, in rain, in rivers, and in the sea their control is obviously a matter of international concern and yet little is known about their movement about the world. The recommendations of the Advisory Committee on Pesticide and Other Toxic Chemicals in 1964 led to a 25 per cent reduction in the quantities used in this country and their more recent recommendations in 1969 will, they believe, lead to a further 20 per cent reduction in the early 'seventies.

109

Naturally, other countries are concerned about the presence of pesticides in the environment, and the extent to which action is taken on this must clearly depend upon their particular problems in agriculture, public health, hygiene, and living conditions. Clearly a ban on DDT in Sweden, Norway and Denmark would be unlikely to affect the health of their people or their economies, whereas the situation in those countries with severe insect-borne disease problems, and where food production is vital, is very different. Nevertheless it would be generally accepted as highly desirable that the quantities of persistent organo-chlorine insecticides used should be reduced as quickly as possible and that governments and industry should endeavour to encourage the development and production of effective and cheap alternatives which it is to be hoped would not also become environmental contaminants. Recently the presence of PCBs (polychlorinated biphenyls), which have been used in the manufacture of lubricants and plastics since the mid-1940s, has been demonstrated[25] in biological material. Initial surveys show that, like the organo-chlorine pesticides, PCBs are found in many freshwater and marine organisms throughout the world in concentrations similar to those of DDT. Great interest and concern in these compounds has been aroused recently because many of the dead seabirds found in the Irish Sea late in 1969 contained considerable quantities of PCBs. Although it has not been proven that these chemicals were the prime cause of death, it must be a matter of some concern as to where such compounds come from, how they enter freshwater and marine organisms, and how they are passed along the food chain.

The work with pesticides and PCBs highlights two problems with which the authorities responsible for controlling pollution are faced. They have first to decide what can be accepted as a safe concentration, for many of these materials have been pronounced safe on the basis of short-term tests and yet, as ecological experience has subsequently shown, they can have long-term adverse effects. The second problem is to discover how many other materials about which relatively little is known are present in the environment and to determine what kind of steps must be taken to identify and control these. The recent reports[26] of high concentrations of mercury found in fish in Canada and Sweden highlight this problem.

It was reported[27] at a recent seminar on Water Pollution by Oil that a serious problem developing on many rivers is the accidental discharge of oil. The incidents that river authorities report are not only thin coloured films of oil but discharges of considerable volume. The marked increase in their incidence is particularly noticeable in heavily populated areas and is apparently directly attributable to the increase in the number of premises using and storing oil. In the area of the Lee Conservancy Catchment Board there has been a fourteenfold increase in the last 10 years in the number of cases of oil pollution reported. In the 1968/69 Annual Report of the Yorkshire Ouse and Hull River Authority there was an appeal for new preventative and legislative measures to deal with this problem.

In conclusion it must be said that most of the rivers in England and Wales contain fish at present. The greater proportion of the total river miles are in good condition and there is some evidence that their condition has improved during the last ten years.[28] However, the volume of sewage effluent and industrial wastes discharged to rivers will continue to increase as the population increases and industry expands. It is necessary therefore to make certain that the requisite technology, expertise and legislation are available and are adequate for river authorities to ensure that the increasing volume of effluents can be safely absorbed into our rivers with no reduction in their quality, use, and amenity value, and that progress towards the goal of "clean" rivers continues unimpeded.

Acknowledgement

Crown copyright. Reproduced by permission of the Controller of HM Stationery Office.

References

[1] KEY, A. (1956). Pollution of Surface Water in Europe, *Bulletin of the World Health Organization*, **14**, 845–948.

[2] HASLER, A. D. (1947). Eutrophication of Lakes by Domestic Sewage, *Ecology*, **28**, 383–95.

[3] KUENTZEL, L. E. (1969). Bacteria, Carbon Dioxide and Algal Blooms, *Journal of the Water Pollution Control Federation*, **41** (10), 1737–47.

[4] VOLLENWEIDER, R. A. (1968). Scientific Fundamentals of the Eutrophication of Lakes and Flowing Waters, with particular reference to Nitrogen and Phosphorus as Factors in Eutrophication. Organisation for Economic Cooperation and Development, Directorate for Scientific Affairs, Paris. DAS/CSI/68.27.

[5] OWENS, M., and WOOD, G. (1968). Some Aspects of the Eutrophication of Water, *Water Research*, **2**, 151–9.

[6] SAWYER, C. N. (1965). Problems of Phosphorus in Water Supplies, *Journal of the American Water Works Association*, **57**, 1431–9.

[7] AWWA Task Group 261OP. (1964). Sources of Nitrogen and Phosphorus in Water Supplies, *Journal of the American Water Works Association*, **59**, 344–66.

[8] OWENS, M. (1970). Nutrient Balances in Rivers: Paper presented at Society for Water Treatment and Examination Symposium on Eutrophication, 24 March 1970.

[9] TOMLINSON, T. E. (1970). Trends in Nitrate Concentrations in English Rivers and Fertilizer Use: Paper presented at Society for Water Treatment and Examination Symposium on Eutrophication, 24 March 1970.

[10] COOKE, G. W., and WILLIAMS, R. J. B. (1970). Losses of Nitrogen and Phosphorus from Agricultural Land: Paper presented at Society for Water Treatment and Examination Symposium on Eutrophication, 24 March 1970.

[11] ANON. (1970). *Chemical and Engineering News*, **48** (16), 35.

[12] BAYLEY, R. W. (1970). Nitrogen and Phosphorus Removal – Methods and Costs: Paper presented at Society for Water Treatment and Examination Symposium on Eutrophication, 25 March 1970.

[13] CROLL, B. T. (1969). Organo-Chlorine Insecticides in Water Part I, *Water Treatment and Examination*, **18**, 255–74.

[14] LOWDEN, G. F., SAUNDERS, C. L., and EDWARDS, R. W. (1969). Organo-Chlorine Insecticides in Water Part II, *Water Treatment and Examination*, **18**, 275–87.

[15] BREIDENBACH, A. W., GUNNERSON, C. G., KAWAHARA, F. K., LICHTENBERG, J. J., and GREEN, R. S. (1967). Chlorinated hydrocarbon pesticides in major river basins, 1957–65, *Public Health Reports*, Washington, **82**, 139.

[16] BAILEY, T. E., and HANNUM, J. R. (1967). Distribution of Pesticides in California, *Journal of the Sanitary Engineering Division, American Society of Civil Engineers*, **93**, 27–43.

[17] ALABASTER, J. S. (WPRL) and LLOYD, R. (MAFF). Private communication.

[18] WURSTER, C. F. (1968). DDT Reduces Photosynthesis by Marine Phytoplankton, *Science*, **159**, 1474–5.

[19] SÖDERGREN, A. (1968). Uptake and Accumulation of C^{14} DDT by *Chlorella* sp. (Chlorophyceae), *Oikos*, **19**, 126–38.

[20] PALMER, C. M., and MALONEY, T. E. (1955). Preliminary screening for Potential Algicides, *The Ohio Journal of Science*, **LV** (1), 1–8.

[21] UKELES, R. (1962). Growth of Pure Cultures of Marine Phytoplankton in the Presence of Toxicants, *Applied Microbiology*, **10**, 532–7.

[22] CHRISTIE, A. E. (1969). Effects of Insecticides on Algae, *Water and Sewage Works*, May 1969, 172–6.

[23] HUANG, J.-C., and GLOYNA, E. F. (1968). Effect of Organo Compounds on Phytosynthetic Oxygenation. I. Chlorophyll Destruction and Suppression of Photosynthetic Oxygen Production, *Water Research*, **2**, 347–66.

[24] VANCE, B. D., and DRUMMOND, W. (1969). Biological Concentration of Pesticides by Algae, *Journal of the American Water Works Association*, **61** (7), 360–2.

[25] MOORE, N. W. (1969). The significance of the persistent organo-chlorine insecticides and the polychlorinated biphenyls, *Journal of the Institute of Biology*, **16** (4), 157–62.

[26] WOBESTER, G., NIELSEN, N. O., DUNLOP, R. H., and ATTON, F. M. (1970). Mercury Concentrations in Tissues of Fish from the Saskatchewan River, *Journal of the Fisheries Research Board of Canada*, **27** (4), 830–4.

[27] TOMS, R. T. (1970). The Threat to Inland Waters from Oil Pollution: Paper presented at Seminar on Pollution. Institute of Petroleum and Institute of Water Pollution Control, Aviemore.

[28] OWENS, M., MARIS, P. J., and ROLLEY, H. L. J. (1970). River quality – timely reassurance?, *New Scientist*, 2 April, 26–7.

Administration in the Control of Chemical and Pesticide Pollution

N. J. NICOLSON, B.Sc., A.R.I.C. (Scientist, Thames Conservancy)

Like Mr Owens, I too am going to confine my remarks to the freshwater environment. Moreover, most of what I say relates to the United Kingdom; the situation is further complicated for large land masses with multi-state river systems.

Because of the interrelation of the various aspects of water pollution, whatever division is made for the purposes of discussion will inevitably lead to a large overlap. Administratively it is not easy to consider chemical and pesticide pollution separately from pollution by industrial waste and sewage. This is my excuse, if any excuse be needed, for straying across my boundaries. My own closing remarks were written several days before I read Mr Drummond's paper.

I would like to begin with the question of pesticide pollution, because it is a topic which has received much publicity, not all of it presented in a rational manner. It is true that the history of pesticide use is not a happy one from the outset. These biologically active substances were used indiscriminately over large areas, especially in the United States where spraying from the air, especially from fixed-wing aircraft, led to localised pollution at relatively high concentrations. Disposal of effluents by the pesticide manufacturers was also less than satisfactory. Accidental spillages such as occurred in the Rhine are almost impossible to prevent except by co-operation and liaison with those people concerned, especially in the encouragement of good practice and the responsible disposal of unused spray and empty containers.

Everyone would agree, I suppose, that it is the long-term effects of persistent pesticides that cause widespread concern. This concern arises out of the bio-concentration of some pesticides, such as DDT, along the food chain, which has led to species of wild life becoming sterile, and in some instances to death itself. As with other catastrophes, lessons have been learned some would say at too large a price. It should be obvious, by now, that new pesticides should not be introduced in future unless much more is known about their long-term effects, especially the possibility of bio-concentration. On the brighter side there is evidence that organo-chlorine pesticide residues are on the decrease. Eventually there will undoubtedly be a range of non-persistent pesticides to suit everyone from the farmer to the conservationist.

Before leaving the question of biologically active chemicals, mention should be made of the potential danger from organic chemicals manufactured, not as pesticides, but for many other uses, and which can turn out to have undesirable biological side-effects. At the moment, the polychlor-biphenyls (PCBs) are under suspicion in this context. PCBs are used as insulating materials, as paints and lacquers, yet have apparently been isolated in relatively high concentrations from the carcases of dead birds. It is necessary, therefore, to widen our areas of scrutiny for the possibility of chemicals which may bio-concentrate in the long term with undesirable ecological results.

Turning to the question of chemical pollution, *what are the problems?*

   (i) The large number and variety of organic chemicals produced in widely varying industries, which can find their way into watercourses by means of effluents or other means. Many of these also inhibit, or tend to inhibit, conventional sewage treatment processes.

  (ii) Many chemicals (including pesticides) have been manufactured for use with all too little knowledge upon their introduction, of their toxicity to animals, birds and fish.

 (iii) A large proportion of these chemicals are non-biodegradable, and may concentrate in animals.

 (iv) These substances are often difficult to detect and identify at the very low concentrations in water that can still cause trouble, e.g. one or two parts per thousand million of phenol in drinking water will produce a harmless but unpleasant taste.

  (v) Last, but by no means least, there is the build-up in our rivers of inorganic salts such as sodium chloride, phosphate and nitrate from the increasing discharge of sewage effluents and changing agricultural practice. For example, it has been calculated that the concentration of chloride in the Thames at Walton will almost double from its present 35 mg/litre in summer flows by the end of the century.

If these problems are considered in the context of a river system whose proportion of effluents is increasing slowly but surely, plus the increasing reliance of large towns and cities on the same rivers for their source of drinking water, the magnitude of the problem becomes apparent.

What is being done?

1. River authorities are becoming more technically able to deal with the advances in technology. By their procedure of consent, in which all discharges direct to watercourses and underground have to conform to an agreed standard of quality, the river authorities attempt to minimise the effect of a given discharge. There is an increasing awareness that consideration of oxygen depletion alone is not enough. But sheer force of numbers and diversity of discharges

means a continuing dependence on the co-operation of industrialists which to their credit is largely forthcoming. A classic example of this in the UK is the agreement reached between the Ministry of Housing and Local Government and the manufacturers of synthetic detergents which has led to a dramatic decrease in the amount of foam on rivers.

2. The water supply authorities, despite their conservatism, are well aware of the gradual deterioration of raw water sources and are having to install more complex water treatment plant.

3. Research is being carried out by many organisations into the treatment of inferior water sources, and on water reuse and reclamation. Even more research of an academic nature is being carried out at the universities into various aspects of the aquatic environment.

4. Much publicity is given to water pollution, albeit much of it of a sensational nature. This publicity is part of the process which paves the way for legislation.

What needs to be done?

It is easy enough to say:

1. Obtain more information on the nature, effect and origins of the chemicals in our watercourses.

2. Give more consideration to the siting of industries, especially those concerned with the manufacture of organic chemicals.

3. Devise more realistic financing arrangements so that there is a real incentive for effluents to be treated to a high standard before discharge.

4. Existing legislation must be brought into a more realistic relationship with the rate of change of development and technology, to deal with (a) control of waste disposal in the ground, (b) changes in agricultural practice, (c) the causes and remedying of unintentional or negligent pollution, and (d) particularly the control of the manufacture of potentially dangerous new chemicals.

5. The social implications of pollution, its consequences and control need to be explained to the people at all age levels, with emphasis on the everyday problems and individual responsibilities as well as the sensational occurrences and possibilities.

But these so-called remedies are, in a sense, only details. We in the United Kingdom are probably the most successful of all developed countries in protecting our environment from chemical and pesticide pollution, though we are not at all perfect. Our organisation in the UK is fairly good, probably in front of the rest of the world. We shall not stay so and continue to set the example we should to the rest of the world unless we reorganise. What is really needed is a co-ordination and integration of all the aquatic activities, a bringing together into a rational plan of the efforts and aspirations of central and local government, river authorities and industry. Any reorganisation carried out should have a proper regard to all factors – not just the simple economics of water supply but all aspects of water use. The same should apply elsewhere in the world, particularly in respect of multi-state rivers.

Only an overall plan can begin to optimise the national or international situation. If such a plan did not make available more finance for research and capital works it could ensure a more effective co-ordination of present effort and resources. Ultimately this co-ordination should be on a global basis to minimise wasteful duplication of effort. Certainly an increase in applied research effort is long overdue.

In conclusion, there is still a temptation to think that problems will be solved by technological advances and a greater financial investment. But these alone are not enough. It is management of the overall problem that is required, and leadership; for it is leadership, not direction, that will save the environment.

Chemical and Pesticide Pollution: the Law

PROFESSOR J. F. GARNER (Professor of Public Law, Nottingham University)

I take it, really, that my job is to say something about the armoury available for those anxious to improve the condition of our rivers, and I am going to concentrate on the armoury of the United Kingdom because I know nothing about the law of other countries. I am also going to concentrate on river pollution. Some of the statutes do apply to the estuaries and I shall have just a little, perhaps, to say about pollution of the sea near the coasts. I will also try and concentrate on what I have been asked to deal with, namely, "Industry and Pesticides", but it is not very easy to separate this from sewage, because the statutes deal with the same thing and do not deal with the three separately or, rather, sewage and industry separately. But I will try not to steal Mr Drummond's thunder too much.

How I have broken my subject up is to talk first on Common Law controls – the historical Common Law of centuries gone by, which is still of some value today, as Mr Drummond points out in his paper, and then, of course, the statutory controls made necessary by what we sometimes call the "welfare state", but I wonder, with what we have been hearing these past two days, if there is much "welfare" about it.

The Common Law controls can be classified, I think, into, perhaps, three: liability for the escape of noxious objects; the careless use of noxious articles or pollutants, whatever we like to call them; and the infringement of property rights in water.

First, then, the escape of dangerous objects, what we lawyers call "the rule in Rylands and Fletcher". This is a rule of absolute liability. If you collect a dangerous object on your land such as a tiger or a large quantity of one of these horrible-sounding objects, such as explosive chemicals, and you allow this to escape, then you are liable for the consequences. You are liable in damages or for an injunction for the consequences proved, provided, of course, that those consequences are the direct consequence of the escape. Applying to our context, there was a case in 1921 where the local authority were using tar on their highway and the tar escaped from the highway on to the plaintiff's water-cress beds, to the undoubted harm of the water-cress, and, of course, the plaintiffs were able to recover damages. Now, if you can show that sort of thing, well and good, but

115

the danger, the difficulty about this particular heading of liability, in our context, is that you are only liable if it is an abnormal use of land, and although this point has not been contested in the courts, I would have thought that the use of pesticides, for example, is so usual these days that it would be rather difficult for the courts to say that a farmer who is using pesticides for the improvement of his crops was making an *abnormal* use of his land. I would have thought that this was, today, a normal use of land, and we know from other case law that what was abnormal in Victorian days is not necessarily abnormal today. So, that heading of liability is not likely to be very helpful.

Then there is the heading of "negligence" or the careless use of a noxious agent; a "dangerous thing", as I used to be told in my student days. I remember my lecturer used to put an inkwell on the top of his desk and say, "That is a dangerous object". And we would all say, "What the hell is he talking about?" And then he would start throwing it around. So, a dangerous object is anything, really, that is used in a dangerous way, if you like, and, clearly, our pesticides are dangerous objects in that sense. Therefore, if you can prove injury, assessable injury, as a consequence of the careless use of pesticides, then, I should have thought, an action in negligence must lie. There was a case in 1926 when an oil company was sending some oil by rail in inadequate containers, or faulty containers, and the oil escaped from the vehicle into some water on a farm, which made the water unfit for the cattle to drink, and the farmer recovered damages for the injury to the cattle. Rather an unusual incident, perhaps, but, clearly, that is the sort of situation. Somebody here has been talking about the washing-out of pesticide containers and that if the effluents from the washing-out gets into something on the plaintiff's property which harms him, then he will be able to recover damages. That is all very well. The difficulty there is that you have got to prove the connection – the causal connection, as we say in the law, and you have also got to have a plaintiff who is prepared to run the risk of not being successful in his litigation, and, as you know, that may be expensive. So this is not a very strong deterrent to people being careless about their use of pesticides.

The third heading is the infringement of property rights. We heard, I think from Lord Hodson yesterday, about the Helsinki Rules for pollution, and we have heard other standards suggested, but the Common Law has much higher standards than any of these. In a case in 1897 in the House of Lords it was said, "Every riparian proprietor is entitled to the water of his stream in its natural flow without sensible diminution or increase and without sensible alteration to its character or quality. Any invasion of this right causing actual damage, or calculated to found a claim which might ripen into an adverse right, entitles the party to the intervention of the Courts." So, if you raise the temperature of the water by 1 degree Fahrenheit (not necessarily even Centigrade), or if you discolour it, or if you put gamma rays, or whatever it is, into the water, then you are certainly offending against the Common Law standard of river pollution, which is much higher than any river authority has ever thought of. Also it is no defence to hold that the water is not already pure, or to say that it has been polluted by Bloggs farther up the stream; that is no defence at all. If you can prove that John Smith has caused some pollution which has resulted in harm to

116

you then you can sue John Smith. But the obvious difficulty is, again, that it rests on the initiative of the plaintiff and less on the proof of the causal connection between the damage he has suffered and the defendant's acts.

I cannot resist telling you about a Canadian case I read when I was thinking about this. It is about a Mrs Hubbs who complained that the drinking water in her well had been polluted by some sewage from her next-door neighbour, and in the first line of the report, the Learned Judge says, "Mrs Hubbs complained that her water was not potable to the District Engineer". Mrs Hubbs managed to recover damages.

It is also a Common Law actionable tort to foul percolating water. To abstract percolating water from the sub-soil is not an offence at Common Law, it is not actionable at Common Law, but if you foul water that another person may want to take, then that is actionable.

So, the Common Law goes quite a long way, but obviously, under modern conditions, this is not anything like enough. Obviously you must have an agency capable of enforcing rules and we must have those rules prescribed and not dependent on the accidents of litigation, on the persistence of a plaintiff and the difficulty of showing, of course, whose action it was and so forth. And so Parliament has developed a quite substantial body of statute law. The most satisfactory part of this – I am sure my river authority colleagues will agree here – are those provisions dealing with discharges to rivers, but I am not going to say very much about these as I do not think they cause a lot of trouble except, possibly, in practice. There are other people much better qualified to talk about that than I am.

The Public Health (Drainage of Trade Premises) Act of 1937 as well as the Public Health Act of 1961, start at source, as it were, and enable the Local Sewage Authority to control discharges coming into their sewers, including discharges of industrial effluent. Then, in turn, the effluent from the sewers, or from industry direct into the river, are controlled by the River Pollution Prevention Acts of 1951 and 1961. Now the river authorities have, I should have thought, as wide a control as they need over both new discharges and existing discharges. The problem here, as I understand it, is not really one of law at all. It is one of inspection, of knowledge, and, of course, of economics. You cannot make an industrialist carry out unreasonable amounts of treatment so as to prevent him from putting anything at all into the river except H_2O, if H_2O ever existed. So that I do not think, as I say, that causes any real trouble. I may be corrected on that but I do not think it raises a legal problem. Where the legal problems arise are on the other statutory provisions which attempt to control other forms of pollution – pollution by the old bicycle thrown into the stream, pollution by spillage, seepage or run-off from the land, and, of course, from pollutants coming from boats on the river. The first line here is Section 2 of the River Pollution Act of 1951. It is made an offence "for any person who causes or knowingly permits to enter a stream any poisonous, noxious or polluting matter. Or if he causes or knowingly permits to enter a stream any matter so as to tend, either directly or in combination with similar acts whether his own or of others, to impede the flow of water in a stream in a manner leading or likely to lead to a substantial aggravation of pollution due to other causes or its

117

consequences." So that if you can show that someone who has thrown an old tin of pesticide into a stream – you know who has done it – and you can show that that is a poisonous, noxious or polluting matter, or is likely to tend to lead to pollution, then there is a criminal offence under Section 2 of the 1951 Act. The difficulty, of course, lies in the many defences. First, a defendant may be able to show it is not a stream. A canal leading to a stream is not always a stream, and some of the tiny rivulets a long way up may still carry the polluting matter a long way down and they may not be part of the main water-course, part of the stream. The legislation does not apply to a lake or a pond, which does not lead to a stream, nor does it apply to some tidal waters. But the biggest defence is "not knowingly" – which turns on the words "not knowingly" – "causes or permits". How easy it is for the defendant to say that he did not know that this pollutant was going into the river, or he did not know that it was noxious – I suppose that is more difficult for him to say – but he did not mean it to go into the river; he is very sorry but maybe he was careless and he did not knowingly permit it to go into the river. There are many cases of such a nature – Mr Woodward was talking to me yesterday and giving me examples: the oil tanker driver, for instance, who goes in and has a cup of tea while a lorry is filling his tank; the pipeline breaks and you get a thousand gallons of oil down the sewer. He did not know, he says, that it was being done and he did not knowingly permit, and in any case he did not know where it would go anyway. And so it has been suggested, and I think quite reasonably, that the law needs tightening in relation to this key section. Perhaps we ought to have an absolute provision; it should be a crime to commit any act which causes river pollution.

The modern attitude to strict offences in the law generally has recently been highlighted by a report of the Law Commission, and the Law Commission have concluded, very tentatively at the moment, that a strict offence is an undesirable feature in the law. The object of the Criminal Law, they say, should be to prevent offences, not to catch the offender and punish him – that is only a secondary point. The primary object of the Criminal Law is to prevent a crime happening. If you have a strict offence, that you are liable whether you know about it or not, whether you mean to do it or not, whether you are careless or not, it just simply happens beyond your control, it just happens to be your tanker where the pipe breaks. In such a case you will not be able to take any precautions against the inevitable accident, and it would seem that a better attitude to this type of offence would be to make it a crime to cause pollution unless you can show a very good defence. In other words – I am not talking in lawyers' language now – the burden of proof should be moved from the prosecution to the defence. The prosecution proves the release of the pollutant and that it has gone into the river, and the defence then has to come up with some reasonable excuse; why they did it, why it happened, and that they could not possibly help it. Now it has been suggested that we should go the whole way and provide for a strict offence and leave the industrialist to insure, but I do not see that this is the right attitude. I do not see that we really want to have pollutants coming into rivers and be able to get the money to put it right again, because sometimes we shall not be able to put it right anyway. What we want to do is to stop the pollutants coming in altogether. To my mind, the strict offence with the burden of proof on the

118

defendant is more likely to make the industrialist careful. That is really, surely, what we should be aiming at. That seems to me a better approach to that particular problem.

Well then, that is Section 2; there also are a number of other sections in the Salmon and Fresh Water Fisheries Acts, 1923 and 1965, which one can really drive horses and carts through if one wishes to. There are some very extraordinary provisions in them. It is an offence "to put into waters containing fish (or any tributaries thereof) any liquid or solid matter to such an extent as to cause the waters to be poisonous or injurious to fish". The difficulty about this is, of course, we do not even know what is meant by "fish". You may do, but the lawyers do not. You see, unless you define these words, lawyers do not know what they mean – is an eel a fish? Crayfish and winkles probably are. Then the Act is enforceable only by the Fisheries Board. There is a defence whereby one can acquire a right to pollute over a long period. Also there is a problem in this particular case of what is meant by "poisonous". "Poisonous or injurious to fish." Is it "injurious" to make the fish, though perfectly healthy, inedible? Is that "injurious" to the fish? I do not know. The fish might be very pleased he is not going to be eaten. What happens under this section is that industrialist A may put in chemical No. 1, which is perfectly harmless to fish, and industrialist B may then put in chemical No. 2, which itself is also harmless to fish, but chemicals 1 and 2, by some odd chemical reaction on each other, when together in the water become harmful to fish. Whom do you prosecute? The man who put it in first or the man who put it in second? It is rather unfortunate if the man who put it in second happens to be lower down the river. Why pick on one? You cannot prosecute both of them but I suppose you have to pick on the second man. It seems a little unfair. So much, I think, for those sections about the adding of deleterious matter otherwise than by a discharge. The control of discharges is the River Authority's job and, I think, something they are able to cope with.

Now, the other little bit of law that exists, and it is a very little bit, is concerned with discharges to percolating waters. There is a section now in the Water Resources Act of 1963 which makes it an offence "to put trade effluents or other poisonous, noxious or polluting matter into wells, bore-holes or pipes, into the underground strata, within a River Authority's area". Well, I am not a chemical engineer and my immediate reaction to that is, "Well, what if he does it by some other method than by a well, a bore-hole or a pipe?" I should have thought he could have let it just seep through, which does not constitute any offence at all.

Control of pesticides themselves: here there is very little legislation indeed. I should have thought that in view of the information we have been given today we ought to be thinking about the control of pesticides, but then the legislation is extremely thin. Presumably most pesticides will be classified as poisons, but if they are not poisons – I do not know whether they are – then there is virtually nothing about them in the legislation. But, assuming they are poisons, under the Poisons Rules of 1952 any listed poison may be obtained by a purchaser engaged in the trade or business of agriculture or horticulture. So there is no difficulty for the farmer getting as much DDT or arsenic to poison all the mothers-in-law he may ever meet. No disadvantage, no control at all!

119

Section 8 of the Protection of Animals Act of 1911 makes it an offence "to place on any land any fluid or edible matter which has been rendered poisonous" – again, this is not defined. But there is a defence; if it is used for the purpose of destroying insects – and sometimes there is a duty to destroy insects such as the colorado beetle and other invertebrates, rats, mice or other ground vermin, where found necessary in the interests of public health, agriculture or the preservation of other animals, domestic or wild, or for the purpose of manuring the land. So you can use poisonous matter to manure the land, "providing reasonable precautions have been taken to prevent injury to dogs, cats, fowls or other domestic animals and wild birds".

It is also possible to use practically any substance for the purpose of killing or taking wild birds if you have a licence from the Home Secretary or the Natural Environment Research Council. I imagine those are given very rarely. It is also permissible to use gas down rabbit holes to prevent damage by rabbits – by the Prevention Act of 1939. It is an offence to use in or near any waters any poison or other noxious substance with intent thereby to take or destroy fish – Salmon and Fresh Water Fisheries Act, 1965, Section 1.

I cannot say I have made a very exhaustive search through the statute book for relevant legislation. I may not have enough imagination as to which sections were relevant anyway, but I think I have got most of them there, and that is about it. That is about as far as the armoury goes. So, to my mind, the law has quite a lot of loopholes. The Common Law is too difficult to operate under modern conditions. I think the control of discharges is, as I said, adequate, but the control over pesticides and, probably, in respect of underground percolating water does need tightening up and there is also this vexed question of what to do with Section 2 of the 1951 Act. I am quite sure it needs strengthening and, of course, also, the penalties need increasing – I think the maximum penalty is £100, although a prosecution can be taken on indictment and then the defendant can be sent to prison. But I am not sure the public would tolerate an industrialist being sent to prison even for putting cyanide in the river. Perhaps if it came before some of Her Majesty's Justices that very laudable result might be achieved.

So that is one thing I think we need doing. I think, possibly, we ought to be thinking a little more about the administration. I am not myself satisfied that we have got the right answer on administration. I would like to hear what Mr Drummond and others have to say about this. It seems to me that we have got too many types of authorities; you have got the sewerage authorities, at the moment very many of them, but, of course, the reform of local government may take care of that; then you have got the river authorities and you also have got the water supply authorities. Do not we need some sort of overall authority for a whole river basin dealing with dirty water, moderate water, and drinking water? I know this raises other problems, but in a report I was reading the other day from the Association of Municipal Corporations, the sewerage authorities were being very bitter about the effect of the conditions imposed by the river authorities on their sewage effluents, saying, "Well, why do not these chaps see how expensive this is to the rates", and so on, and so forth. I do not think this kind of consideration should come into it at all. We should look at it as an environment

120

pollution problem with one authority for one water-shed area. Well, that is only just a suggestion.

Discussion

Dr D. W. Bowett, Chairman

Could we perhaps start from the beginning and see whether everyone is satisfied that there is a case made out for tighter regulation and control. Is there anyone here who doubts the magnitude of the problem?

Mr J. C. Hanbury

I come from the chemical industry and I am also a member of a river authority, so I am a poacher and gamekeeper rolled up in one. I was very interested in hearing Professor Garner's outline of the legal situation and I had the pleasure of a long conversation at lunch yesterday with him about this, but may I just fill in one gap, because, unless the gap is filled, people may be under a mis-apprehension. In regard to the present situation on pesticides, agricultural chemicals and potentially harmful materials used largely in agriculture, there are, I know, in this room, people from the chemical industry more directly concerned in these things than I am. There are also people from the Ministry of Agriculture, all of whom know more about it than I do, but I would like to say that whatever the strict letter of the law may require in respect of agricultural chemicals, there is nowadays an exceedingly close working together between industry and the Ministry of Agriculture. I was under the impression that this was enshrined in certain legal sanctions of a rather vague kind – I do not know quite what they are – but whether one can be brought to court or not, the close working between the Ministry and the industry is such that some of the shocking things that have happened in the last fifteen or twenty years are, with every day that passes, increasingly improbable. There is now a booklet, I think produced by HM Stationery Office – a red book, some of you may know it and I am sorry I have not got it in my hand at this moment – in which all these things are written down with great precision. It is full of information about the industry that produces agricultural chemicals and about the Ministry of Agriculture that sanctions their use subject to appropriate safeguards. These are first of all to protect the health and persons of those who will handle them, very often not very well tutored farm labourers. One must know the toxicology of these things so that one knows what the hazards are on the medical side, and then there are the veterinary aspects, hazards to livestock, the sort of mishaps that can occur and how serious they may or may not be, and then ecological considerations, the proper manner in which they must be used, the pitfalls that must be avoided, the things you must not do. All these things are enshrined in writing in documents. I have no doubt Mr Owens knows more about this than I do, but I just wanted to say that there is not negligence on this subject, and I hope that others, better informed than I am, will, perhaps, bring out some of these points more precisely than I have done.

Dr D. W. Bowett, Chairman

Thank you, that is very helpful. It does cut at a tangent to one of the points made by Mr Nicolson. The point was that we do not have the kind of control of

121

chemical and pesticide manufactures that we have over the manufacture of drugs. That is to say, there does not have to be a prior permission for the manufacture of these things as there does have to be for drugs. So perhaps Mr Nicolson could reply to that in due course.

MR A. S. WISDOM

I would like to deal with one or two remarks made by Professor Garner while they are still fresh in my mind. Professor Garner was very comprehensive in detailing the cover of most of the law; I am not certain but that the best course for the other lawyers present is to pack up and go home now. Be that as it may, may I just make one or two comments on certain aspects he dealt with that he did not necessarily cover entirely, because nobody can cover all the law involved in the brief space of twenty minutes or so. He referred, in particular, to one of the Common Law controls, the infringement of property rights – by that he implied riparian rights. Well, of course, really there is nothing left worth talking about of what we know as riparian rights – that is, the right to abstract, the right to impound water from a river and the right to quality, because they are now all dealt with under statute. If you want to abstract water, you have got to get a licence to abstract, subject to certain exceptions. If you want to impound water from a river, you have got to get a licence, again, from the River Authority, and you have got to get consent under Land Drainage. Likewise, if you want to discharge water into a river – effluent of any kind – again, you have to get the consent of the Authority, and the same goes for discharges into underground strata. So, really, very little is left of what we used to know as riparian rights, and about which professors, such as Professor Garner, were able to write comprehensive textbooks covering several hundred pages. He then went on to deal with Statute Law, and I would not want you to think that everything is easy-going there. We have got the controls of the River Authority, we can deal with offences, but, of course, that is not the whole picture and Professor Garner did deal with instances where it was difficult to enforce the law in a bad case of pollution. Now, first of all, what is meant by pollution? A lawyer cannot define it, he has usually to get a chemist or a biologist to do that for him – it is a team effort – and very often it is difficult to know when pollution has been caused, or is likely to be caused – that is one difficult problem. Then, the second thing, what is the standard of purity you look for in a river when you suspect that pollution has been caused? Nowhere in this country have there been laid down any legal standards of what is meant by pollution, except under the Oil in Navigable Waters Acts, which was dealt with to some extent yesterday – how many parts per million of oil in mixture – but, apart from that, there is absolutely nothing. Fortunately, most courts, at least all courts I have been before, will accept what is known as the Royal Commission Standard, or whatever other special standard is necessary when the Royal Commission Standard will not do, but, of course, you have got to prove it – you have got to bring your chemist into court for him to show them the sample, to say that he analysed the sample and that his report was that the river was grossly polluted, or whatever term he cares to employ. Then we have trouble, as Professor Garner stated, with trying to enforce Section 2. It is very much of a gamble at the present time

122

whether a really bad case of pollution can be taken under Section 2 of the Rivers (Prevention of Pollution) Act, 1951. Certainly anything in the nature of accidental pollution, or pollution outside the control of the person who caused it, cannot usually be prosecuted. If you take a prosecution under those circumstances then you are liable to have your case thrown out and perhaps costs awarded against you and, therefore, that is one of the reasons why river authorities cannot bring as many prosecutions as they would like under Section 2, prosecutions which, in practice, they perhaps should have brought – they have missed the boat on many occasions. And then there was this reference to Section 72 of the Water Resources Act, which enables the River Authority to grant consent to discharges made underground, but this only, of course, applies where it is into a well, pipe or bore, anything outside that is not controlled; if you put water into a natural fissure, then, of course, you can get away with it. Whether a disused coal-shaft amounts to a well is unclear though I think possibly it does, but that sort of discharge ought to be brought under control. I think, sir, if I may just sum up, whatever controls you have, or whatever controls you want to have, you will never get rid of pollution and the unforeseeable pollution which can never be controlled. Whatever the laws or whatever tests you have for it, there will always be pollution of that kind.

Dr D. W. Bowett, Chairman
I would like to ask members of the panel if they would like to reply to a specific comment.

Professor J. F. Garner
I would like to reply to just two small points. In the first place, Mr Hanbury and his liaison between the Ministry and industry – I was not aware that it was as close as this and it is certainly a good idea, but I am not at all sure that this really meets the case of the careless farmer, I mean the careless farmer who cannot, or will not, read the nice books put out by the Ministry. I am not sure we have got enough control over him in the legislation. And then, Mr Wisdom, who really knows very much more about this than I do, and has certainly written a lot more than I have on it. He makes the point that the Common Law riparian rights have almost gone. I admit that I did not refer to the Water Resources Act, but I do not quite agree with him, with respect, that they have almost gone, because, after all, the River Authority does not control the very small measure of abstraction of water caused by the watering of cattle on a river bank, and this may be the case where the farmer wants to use his Common Law rights. He has a Common Law right to use the water coming past his land for natural purposes and the Common Law has always thought that the watering of cattle is a natural purpose.

Dr J. A. G. Taylor (Unilever)
As you know we are a large firm that produces foodstuffs. Taking up Mr Hanbury's point about the use of pesticides, etc.: because we contract for much of our foodstuffs from farms we are able to direct how these materials are used and have a guide to ensure that they are only dosed at the right level and at the right time. Furthermore, we rigorously monitor the raw materials we sub-

sequently purchase. So you can see that in these areas industry is again showing a responsible attitude towards potential risks. You take out a relevant segment and say, "This is what you must do with your particular crop, with your particular field. You must put it on in this way, and to that amount". We, in fact, have control to make sure they do use these techniques. I think the relevant thing is that you refuse the product if it has too much this, that or the other in it. If there is too much penicillin in the milk, it is rejected. Because we have a contract situation, we can say, "We will buy from you, if you do it our way".

MR F. MacDonald

Normally, a manufacturer is not the primary user of these insecticides – there is someone else who is involved. Now, industry is getting very concerned about its reputation in these environmental controls. They are now embarking on a very large programme of education of the end-users, to such an extent that it is mandatory, in some cases, that they will not sell their products to certain industries who show irresponsible control of these products. But this does not prevent an industry from importing this product from Germany, France or somewhere else. This is an extremely difficult situation to get into – we actually lose business through this, and I think this is something which is really of concern and must be tackled.

MR E. H. Hubbard

I want to speak in support of the last two speakers as a member of an Oil and Water Working Group in England – I am also a member of an Oil and Water Working Group in Holland, and two of my colleagues are members of one in Germany. I think we must distinguish – I do not think the law does distinguish – between the regular operations of industry, which are easy to control, and the accident. As the last speaker said, a product is sold to a wholesaler and he sells it to a lot of distributors who finally sell it to the customer. At every step there is transport, handling and storage and sooner or later, someone has an accident. When an accident happens everyone remembers the product but not the person who caused the accident.

One of the problems which certainly occurs in England is the question of pollution of rivers by small oil consumers. I happen to have a friend in the water industry who looks after the River Lea, where there are a large number of market gardeners, and he is always having this type of trouble.

A true but ridiculous example was the case of an oil tank with a pipe to the burner for heating a greenhouse. A thief came in the night and stole the brass connection because he wanted to sell the brass. The next morning there were 2,000 gallons in the river, and the first intimation my friend had was when he was telephoned by the owner of a yacht marina who asked who was going to pay to have the yachts cleaned. What we do have to try and do is to distinguish between a regular operation and an accident. It is true that industry has accidents and we must not hide the fact, but industry also has the means to clear up the mess quickly. It is more often the user or the transporter who is involved in the accident, however, who may not know how to clean up the mess, and often does not report it.

MR J. R. BUCKENHAM

There are two points that I should like to make, one raised by Professor Garner, and one raised by Mr Hubbard. Mr Hubbard first of all: I would like to support what he said, and add to what he said about accidents. My Association, the British Waterworks Association, representing water supply undertakings and the Association of River Authorities, are both very worried indeed about the increasing incidence of accidental pollution, and both organisations feel that more could be done to cut down the category of avoidable accident, in which there have been some pretty hair-raising incidents over the last year or two. One such which comes to mind immediately is the overturning of a very large road tanker in the north of England, filled with some fluid so complex that not even the manufacturers had any clear idea of what its composition was. But it was all over the road, and along came the fire brigade, and the River Authority looked on with horror as the firemen enthusiastically squirted their hoses on the road and washed the fluid all down the nearest water-course, whereupon it slaughtered all the fish downstream for ten miles. This is the kind of thing which, if it happened alongside a road with a ditch, which eventually led into a river or other source of supply for drinking water, could cause a good deal of alarm and despondency, to say the least. This incident highlighted a number of things; first of all the structure of the tanker; secondly, the desirability of hauling this stuff, whatever it was, around the countryside in such a tanker; thirdly, the rather precipitate action, though well-meaning no doubt, of the emergency services who thoroughly flushed the road and got rid of it all, to their satisfaction, thereby handing it all to the River Authority; and finally – the unfortunate driver having no idea of what was in the tanker, or whether it was toxic or not – the tanker should have been properly labelled. Now, a lot of these things are being investigated. The tankers now have to be labelled, and their construction is something that the Government are looking at in a very detailed way. The emergency services, too, are a little more educated about it and are consulting us and our river authority colleagues before the stuff is washed away, but this incident, among others, does I think illustrate the dangers of avoidable accidental pollutions and the necessity, too, of something being done in all those fields to ensure that these increasingly sophisticated chemicals, in increasing volumes, are not allowed to get into a situation where this sort of thing can happen.

Another case, which is nearer home to Mr Hubbard, of course, is the question of pipelines. There are many pipelines now which criss-cross the country and which are being used for the transport of all kinds of materials. Many of them cross rivers and countryside where there are water-bearing rocks in the sub-strata. Even bursts which are minor from the point of view of the operator can be quite disastrous to an aquifer, say, chalk in the South Downs, from which water is being extracted. There again, on the water supply side, one looks with apprehension at what advances in technology are bringing us in the way of further, and bigger, perils. I am merely saying this in order to show that accidents of the kind that Mr Hubbard referred to really ought to be reduced to the absolute minimum by manufacturers, by industrialists, and by others who have an interest in these things, because they can be very frightening indeed when they happen.

I think that is enough about accidents, Mr Chairman, but I would like now to turn to what Professor Garner said. He threw out, casually, at the end of his paper, a suggestion for what is now being termed "Integrated River-basin Management". This, as I understand it, means that there ought to be one big body which concerns itself with water in its natural state, takes it out, purifies it, supplies it to the customers, takes it away when it is used, cleans it up again, and puts it back in the river. This is a good idea on paper, although nobody has suggested yet that coal merchants should be responsible for hauling away ashes. However, I cannot help feeling, Mr Chairman, that there are two things about this. First of all, where you have one integrated authority which concerns itself with looking after the purity of river and effluents, and extracting water, and supplying it to people, and removing the used water, you remove the element of the "policeman" which the River Authority exercises at present – an independent body looking, with a very beady eye, at people who put stuff into rivers; they are in a position where they can monitor these things. If they become integrated with the people who are busily putting stuff into rivers from public sewage plants, and with the water supply authorities, I cannot help feeling that their position as "policemen" is weakened beyond the point where it is even today, where some think it is perhaps not strong enough as it is. I cannot help feeling, perhaps a little unkindly, that what is behind Professor Garner's suggestion is the thought that of all these groups of authorities – the river authorities, the sewage authorities and the water supply undertakings – only the latter are the ones who pay their way without any kind of government subvention, and I cannot help feeling, Mr Chairman, that it is our money they are after.

DR D. W. BOWETT, CHAIRMAN

Do you think we could pursue this point, because it is clearly important – the whole question as to whether you are to rationalise the management of a river-basin, and somehow integrate or co-ordinate the various authorities. We will just stick with this one, because it really is important, for a few moments, and then I do not want to suppress free speech any further.

PROFESSOR P. F. WAREING

I think that, probably, Mr Drummond will have something to say on the following matter, but I would like to make one point, namely, that those who are advocating the control of all operations relating to the use and conservation of water on a river-basin basis are not proposing that the supply undertakings should necessarily come under the same authority, but the argument is, rather, that you cannot separate water quality from water conservation. Take, for example, the River Trent. The Trent could, in fact, supply a large part of the water requirement of that part of the country if it was not so heavily polluted. Since water quality and water conservation are so intimately connected, it has been argued that we should take the river basin as the basic unit and have unified control of both sewage treatment and water conservation activities, but not actually the house-to-house supply. With regard to this question of "policing", it would not require a great deal of imagination to have some other agency, if this is necessary, to check on the water quality problem.

126

SIR FREDERICK WARNER
I was going to say, Mr Chairman, that we shall get more light on this matter
once Mr Drummond has spoken, because there is not only the proposal that the
river authorities should take over the whole of the integrated bodies. I think it
has been mentioned there is the proposal from the Association of Municipal
Corporations that in the reforms of local government, whether they follow
Maude or not, these powers should be taken over by the local authorities, and I
think you will find that there will be a big conflict of interests arising immediately
when we start discussing who should exercise integrated control. I think, again,
Mr Drummond's paper will show us some of the problems in exercising
integrated control because, in the basins, we are very far from having any
workable models which enable us to decide what is the effect of our action in
any one place and where the money is best expended. As he says, we are carrying
out a study which will give some pointers in this direction. But this is only the
first such step. We do not have this in many other parts of the country.

DR D. W. BOWETT, CHAIRMAN
On the assumption that Mr Drummond is going to tackle this, perhaps we ought
to leave it until this afternoon.

MR J. R. STRIBLING (Thames Conservancy)
May I ask Professor Garner if he would like to be so bold as to give us his
definition of the term "underground strata". Now, as a practising Pollution
Prevention Officer myself, I well remember in 1965, when, at that time, one of
my superiors, to wit my friend Mr Wisdom, said, "Well, the Water Resources
Act has come into force. Here is a copy of the Act. Get on with it. Do not ask
me what to do, I do not know any more about it than you do." And when I got in
touch with him a few days later, and asked him what his interpretation of the
definition of "underground strata" was, he whispered something about having
an urgent appointment. Subsequent to that two or three of us went to the
Ministry and we had a long interview with Dr Key who has since retired, and
I asked him if he could tell us what he thought was in the mind of the Ministry
when that section of the Act was drafted. He gave us to understand that the
Ministry considered that there were so many septic tanks in the country whose
job was to discharge the overflow via sub-irrigation drains, and other means,
into the soil, and that these were working quite satisfactorily, and that to say
to the householder, "You shall not henceforth so discharge, you must either
convert your septic tank into a watertight cesspool with all the tiresome frequency
of emptying, or you will have to put in a small purification plant", the Minister
thought that would bring about severe hardship on the owners of those houses.
Therefore, it was put to me that the Ministry considered that, since most of
these overflows took place in the top 4 ft of soil, it was the proper heritage of
people to so use that top 4 ft, and that, therefore, the underground strata was
likely to be construed as starting from 4 ft downwards. The only flaw in that
argument, of course, is that where the top 4 ft or any part of it consists of chalk
or an aquifer, then obviously this could not be permitted, and so, ever since 1965,
in operating Section 72, since I have not heard from anybody a better definition
of "underground strata", I concluded that it should commence at a depth below

4 ft (with the exception of chalk). I have operated it on that system ever since and I would be very glad to hear whether Professor Garner has any views on the matter.

PROFESSOR J. F. GARNER
I would only say, Mr Chairman, that "underground strata" is wherever the geologist tells us it is underground. A little while ago Lord Denning is on record as having said that a highway goes down – you know the surface of the highway belongs to the Highway Authority – goes down to "two spits". The chap spits on his hands, digs his soil up, chucks it out, has another spit, and chucks another lot. I think it is a comparable question. It is just as far down as it needs to go down. That is really as far as I can go until the courts tell me something better.

LORD SIMON
I only wanted really to ask one question of Mr Nicolson. He rather threw away a remark in his interesting talk indicating that he thought there would soon be non-persistent pesticides, and I really wondered whether any of our scientific friends here could say how people are getting on with it, because, if you have got non-persistent pesticides that are easily broken down, well, some of the problems discussed will become immediately much less serious. You can have the same situation we have with detergents. When you once have degradable detergents, then you can prohibit the use of others.

DR J. W. HOPTON (Birmingham University)
I suggest that over-legislation can perhaps inhibit the solution of some of these pollution problems. Certainly there is evidence that the adoption of eminently sensible standards can have this unfortunate effect. Bacteriologists like Frankland, at the end of the last century, demonstrated the great value of the Presumptive Coliform Test for the assessment of faecal pollution, but paradoxically their success had an adverse effect on research in aquatic microbiology. This was a pity because these early microbiologists were clearly interested in the total microbial flora of waters but much of the work in aquatic microbiology carried out during the first forty years of this century was concerned with the measurement and persistence of a few species of pathogenic bacteria. These bacteria are important medically but their contribution to the metabolic capacity of aquatic systems is negligible compared with that of other bacterial species.

Moreover, perhaps tight legislation would discourage the application of knowledge which has accumulated from pure research. One of the most interesting things to come out of post-war research in microbiology has been the finding that the metabolic pattern of micro-organisms is very much influenced by the chemical and physical condition of the environment. The activity of enzymes can be profoundly affected by the presence or absence of, at first sight, unrelated chemical substances. If one fixes, rigidly, the permissible level of, say, an insecticide in an effluent, is it possible that people will leave it at that and be disinclined to investigate the prospects of achieving complete detoxification by controlled microbiological action?

MR F. MACDONALD
On the organo-chlorides there is research being carried out now, and what they

are identifying are their side effects. There are certain impurities, which are the real problem, and it is too early yet to say how far they are going to be successful in getting rid of or dealing with them. But they have started this work and it is these tiny impurities in the organo-chloride compounds which, in some cases, prevent the biodegradable action which can follow their use.

Mr A. J. O'Sullivan

In answer to Lord Simon's question, I understand that there are a far greater number of biodegradable herbicides developed than biodegradable insecticides; the only one of the latter that springs to mind is the natural product pyrethrum. I have a feeling that the future of insect and pest control may come to depend more and more on biological weapons rather than chemical ones. Chemical insecticides are relatively crude and are not species specific; that is, they kill beneficial insects as well as pests. Furthermore we cannot even assume that the extermination of pests leads to any increase in food supplies for man. Pesticides become concentrated as they go through the food chain, the predators also get killed and as they breed more slowly than their insect prey, a proliferation of the insects results. This can be dealt with only by further application of pesticides, thus bringing about an escalating situation.

Biological control, on the other hand, has a far greater chance to reduce greatly and permanently the population densities of many pests and does not reduce the diversity of the ecosystem in the same way as chemical pesticides do. For example, one of the methods used has been the release of irradiated male insects of the pest species which then mate with the females which in turn release infertile eggs. This method has been used to completely eliminate the population of the screw worm on the island of Curaçao. Insects themselves can be used to control other pests. Prickly pears (*Opuntia* species) once occupied about 50 million acres of land in Australia; the introduction from South America of a caterpillar, *Cactoblastis cactorum*, which burrows into the prickly pear plant, provided substantial control of the two main pest species over most of their range. Other methods of biological control which have been successfully used are based on the sex attractants which a number of insect species use to bring the two sexes together for mating. These sex attractant substances are highly specific for particular insects; once their structure has been discovered they can usually be synthesised and can be laid in microscopic amounts in the affected areas. The pest insect can then be lured to a poisoned container or its behaviour can be so grossly altered that normal mating does not take place.

I would like to add a small comment to Professor Garner's very comprehensive paper. My own work is connected with the protection of sea fisheries and with the control of pollution in coastal waters. One of the most useful pieces of legislation we have, and in fact the only legislation that is capable of restricting pollution of the marine environment, is the Sea Fisheries Regulation Act of 1966 which incorporates an earlier Act of 1888. This gives local Sea Fisheries Committees the power to regulate and control the discharge of substances detrimental to sea fish or sea fishing. There is some difficulty in defining "detrimental" but perhaps not as great as the difficulty Professor Garner mentioned in defining "injurious". The Association of Sea Fisheries Committees is at present con-

129

sidering a revised form of by-law in which it is hoped to make this definition clearer; here "detrimental" is defined not by means of a list of special substances which would become rapidly out of date, but in terms of the effects of the substances on the fish population and the receiving waters. Furthermore the revised by-law takes into consideration effects other than those acting directly on the fish population; for example, destruction of habitat, spawning grounds or food organisms are considered detrimental. If this by-law is of interest I can give more details of it this afternoon.

Dr J. R. Goldsmith

I would like to speak first to the question raised by Mr Nicolson, concerning the proposal that new pesticides and substances should not be introduced unless it is proven that they produce no harm. I fully subscribe to this in principle, but to put it into operation requires a fund of knowledge which is often unavailable, or has not been collected or does not exist. I think it is necessary not to contest the principle, but to study the mechanisms by which it can be applied efficiently. One of the mechanisms is to ensure that what is learned by someone's misadventure is widely enough disseminated, thoroughly enough studied, and applied in future operations so that the misadventure does not need to be repeated. In this regard I find it somewhat disappointing that there has been no mention so far – or if so I have missed it – of one of the most serious problems for human health through pollution, which has to do with mercury. Here, epidemic disease of the central nervous system has been produced in populations in Japan, and a number of fatalities have occurred. So far, as far as I know, with the exception of Sweden, there has been no serious effort on the part of authorities to find out what the distribution of mercury is in water, and in fish which concentrate it from water, in order to make sure there is no hazard or, if there is a hazard, it is of no consequence, or adequately controlled. It seems to me a matter of pre-eminent importance, and I think it, perhaps, deserves more attention than it has received since these compounds have been used rather widely as fungicides for the treatment of seeds. Most of the acute problems of toxicity have occurred with more heavily polluting industries, many of which are now limiting their use of this substance – I think that is very creditable.

I also want to speak briefly for the type of function the World Health Organisation, which I am partly representing here, carries on. It collects information and disseminates it, it undertakes to have information about pollutants thoroughly reviewed by scientific authorities and, when appropriate, it sees to it that the information is incorporated into environmental quality criteria. The Environmental Pollution Unit of the Organisation is working on air quality criteria now and the Community Water Supply Programme has published international standards as well as participating in the establishment of European drinking water standards.

In many cases, these standards themselves need further improvement, further refinement. I think any agency that is concerned with the standard-setting procedure has first to make sure that there is an adequate input of valid information, of which information about human health effects is of great importance – I am afraid it has not been very much mentioned here – but it also must make sure that there is scientific review by authorities with international influence and regard, and also that when new information is obtained, that information is promptly applied to these standards.

We at the moment in California are studying another problem of water pollution which has only been mentioned in passing, and that is the problem of the pollution of waters by nitrate, which leads to the possibility of the disease methaemaglobinaemia in infants. We are attempting to see under what conditions this occurs as a result of using water from an already mineralised underground aquifer. The pollution of underground waters fully deserves, also, a great deal more attention that it has received, because, once it occurs, there are very complex conditions concerning its prevention and the restoration of the quality to underground waters, and the trend to water re-use certainly is likely to aggravate the risk to polluting underground water supplies. There are three groups in three different countries that are now studying this particular problem; a group in Israel, where water re-use is a very urgent matter, and it is proceeding very vigorously; a group in Czechoslovakia, where the natural mineralisation appears to be the cause of the high nitrate level; and a group in California, which I am responsible for, in which it seems most likely the cause is rather complex and has to do with the attachment somehow of nitrate to soil particles. With the rise and fall of the aquifer, this becomes apparently re-dissolved. Elevated ground-water nitrate is not apparently an immediate consequence of the use of excessive nitrogen fertilisers, but maybe a delayed one. With respect to Professor Eisner's comment, the residence time or the delay time would have been utterly unknown except for this occurrence and certainly is not very well understood even now.

Mr A. H. Walters

Regarding the point on the relation between nitrate run-off and over-chemicalisation of the soil, I have recently published a review of the world's scientific literature on nitrate in soil, plants and animals. This shows how excess quantities of nitrate applied to soil can be taken up into plants and so into animals and possibly humans. At our own farm we have been watching what happens when, under similar conditions of overchemicalisation, rainfall causes washout of nitrate and run-off into ditches. There is no doubt that overuse of chemicals in modern farming is causing pollution of soil, water and plants. Professor Garner has coined for us the marvellous phrase – "normality in abnormality". The abnormality of modern farming is now considered normal. Where do we draw the line? I feel this is really Dr Hoffmann's point. From the lawyer's viewpoint it would seem that a lot more carefully documented information from world sources is needed before this question can be answered, and I would be happy to supply some of the information.

Mr G. W. Hull

I should like to link Lord Simon's question with the remarks of several recent speakers. If one continues to push into the environment new things or old things in increasing amounts, there will be a change in the substrata – in the food – of the biological base of micro-organisms that supports in the long run all senior predators of which man is the chief. If we tinker with this dynamic equilibrium, whether with ions of nitrate, phosphate, fluoride, or detergents whether biologically degradable or not, we shall shift the balance. We are, in fact, tinkering already with evolution. So we need to get this considered at the highest level of policy-making in all countries. Professor Garner has indicated some of the legal means by which technological intervention with environment can be ameliorated, but he has pointed out some weaknesses, notably in the Common Law. He has also suggested that we ought to examine educational policy; this must be the final answer but it is a long-term answer. The urgent thing is education and influence upon those concerned with policy planning to get vital short-term steps taken, and this involves also immediate influence upon the instruments of economic planning, and at highest level, if we are to avoid the dangers spoken about during the last two days here. I feel we should address ourselves during the time left to suggesting how this influence can be brought about, and to what specific suggestions can be made.

Dr J. A. G. Taylor

Can I just put the record straight on the comment about mercury poisoning in Japan? Some of you may get the impression that it is from general usage, but what I think was being referred to was a particular incident, that of a manufacturer who was using mercury as a catalyst in a production process. He was not taking sufficient care with his effluent. It was discharged into a river, it was concentrated by the fish, the fish were eaten by the local inhabitants, and this is where deaths occurred. Now, this is not to say that mercury is not something we should concern ourselves about; the fatalities in Japan you referred to are just this one specific case, and it is a horrible disease – I have seen the film showing the effects of this. It is horrible, not just for the fatalities caused; the results are something not nice to see.

Dr G. Howells

I would like to support Mr O'Sullivan's request that what is to be defined as injurious or deleterious should be related to the biological effects. One of the difficulties was already indicated by our second speaker this morning who said that one or two parts per million of phenol in water could be tasted but could not be measured practically. There is exactly the same problem with pesticides, notwithstanding the figures used. Since for the most part the levels in water, which we know have led to biological accumulation and are biologically recognised as injurious, are levels which it is almost impossible for ordinary water quality laboratories to measure. On the one hand, I think legislation has to take into account the biological effects; on the other, we need to promote a better technology which, of course, we already have in the radiation field, and that is why, perhaps, legislation there is more effective than in the general field of water pollution.

Dr A. J. Holding

I think it will surprise many people if on further investigation the increasing use of nitrogen fertilisers does not lead to increased nitrate pollution of rivers, lakes, etc. I would like to raise a general question on the control of nitrate pollution from agricultural sources. We are dealing here with 300,000 or so farms where nitrate fertilisers are being used to carry out good crop husbandry practices. Farmers want to obtain as high yields as possible and manufacturers wish to sell as much fertiliser as possible. Who should accept the responsibility for the control of this kind of pollution? Is the onus to be entirely on the farming community or perhaps the fertiliser manufacturers might become more involved in pollution control?

Dr Goldsmith has referred to the serious problem of ground-water pollution by nitrate in the United States. I understand that in certain restricted areas of Essex, Lincolnshire and Kent the nitrate content of water is causing concern. The intensification of agriculture, particularly in small water catchment areas, could increase the nitrate content quite quickly. I would subscribe to the view that we should be paying far more attention to this point.

Dr D. W. Bowett
It was, I think, Mr Owen's point that there was no known correlation between the use of fertilisers and the incidence of nitrates in the soil.

Dr A. J. Holding
I think it will surprise many people if, on reflection, there is not a correlation between the two.

Sir Frederick Warner
The point I have comes from the last point that has been raised, and refers back to the general point raised earlier. We often have these arguments about the effect of added nitrate, and nobody refers to the fact that if you pile the surface of the soil with dung 2 ft high, in order to avoid using synthetic nitrates, you will still have the activity of clover and everything else which fixes perfectly natural nitrogen in the soil in the form of nitrate. We have always had nitrates in the soil, we would have had no agriculture without nitrates in the soil, and they occur naturally. Most of the problems we have occur as we want to produce more. You can have non-persistent insecticides if you want to go back to pyrethrum and nicotine – they will cost you more; they will be less effective. They might do a lot more for the economies of some underdeveloped countries. The problem for you to decide is whether you want increasing populations, what sort of standard of living you want, whether you want to help underdeveloped countries, or whether you want the biologists to apply something of the technique of the screw-worm fly to the human population. But, to come back to your original point. We have a general point here on which many of us who are scientists, engineers and administrators would welcome the guidance of lawyers and politicians. It arises from the fact that, when we hear discussions of the legal sanctions which are needed, nobody really clears up for us the point at which we have the interaction of politics and law with the activities of the ordinary man, and individual responsibility. I always have the feeling that by too much statute we throw out the baby with the polluted bath-water. In other words, if we look at the areas in which we have made what I think are the great advances in Great Britain in the field of pollution control, they have come about by voluntary means, by the operation of Common Law, by the good sense of Her Majesty's

Judges in making new decisions, and not really by statutory control. The Alkali Act, for example, in England, has produced a purer air than in any other industrial country of the world, by the operation of an Act which includes the phrase that "the best practicable means shall be employed", and this has given no trouble – it has given rise to very few prosecutions in one hundred years. If you take another field in which there has been a great improvement, water pollution, there is the question of the effect of synthetic detergents. A committee of the Ministry of Housing and Local Government (and one of its members is here, Dr Windle Taylor) encouraged research into how you can make detergents biodegradable. They have carried on by voluntary action to control the problem of foaming in sewage works. (Actually, this is another of your Common Law cases besides the Pride of Derby Angling Association. Lord Brockett, if I remember correctly, obtained an injunction against Luton Corporation for causing foam in the River Lee as it passed his estate.) The activity of the Detergents Committee has satisfactorily dealt with the problem of the domestic detergent, and I am quite sure it will be capable of dealing with the problems of nitrilotriacetic acid, if that comes into common use or is intended to come into common use. I am very anxious that we do not lose what we seem to have in this country, which is the ability, perhaps because of the Common Law basis that we have, to make progress continuously. It is the kind of argument that you always hear from Lord Denning, which impresses me as a layman, that you make progress because you do not have rigid controls, do not have rigid limits fixed, and therefore are able to progress and be ahead of time.

PROFESSOR E. EISNER
I would like very much to pursue this question of what protection the Common Law affords. I have been astonished to learn at this conference just how effective the Common Law might be. But it clearly is not being used, and I think one of the things we ought to ask ourselves is, "Is there anything that can be done systematically about this?" Is there any thought of bringing together associations that might enable people to take up their rights under Common Law, where, at the moment, they either do not know them or else have not the means to pursue the matter in the courts? It does seem that this method has the versatility that no statute could possibly provide. I am rather disturbed to see in Mr Drummond's paper this afternoon the statement, "Clearly, in a modern society, riparian rights are impracticable, and will have to be severely curtailed". I think that deserves to be discussed this afternoon when Mr Drummond speaks. I think there is another point that is related here and is very important and that is that we ought to be more realistic in talking about the costs of prevention of pollution. All these things are tied together, and I hope we are not going to hear again the statement that, of course, we shall have to pay for this or that prevention of pollution. Mostly we do not know what the cost of preventing pollution is, we know only one side of the ledger, and we do not know the other. It is more likely than not that in most cases the cost of preventing pollution is negative. I think, if one looks back at examples where these things have been done in the past, one often finds that even in the narrow sense, even in the industrialist's own account, it eventually comes out as a credit balance. Apart from anything else, indus-

134

trialists will find it more and more difficult to recruit people to work in areas that they have themselves polluted. I think we ought to stop saying that it is expensive to prevent pollution until we have some facts to work on. The fact that it costs so much to put in a filter, or some processing plant, is irrelevant at the moment, until you know what the other side of the ledger is.

The last point I would like to make concerns accidents. I do not think we should regard accidents as hopeless. I think again that here the manufacturer has some responsibility to make sure that he supplies the material which he knows is going into untrained hands, in a form in which it is less likely to be misused than it so often is at the moment. We have examples of this already in the medical field where the manufacturers – I do not know whether it is required in law, but as a matter of practice – supply their materials in forms that are not as readily misusable as some of the things we have heard of are.

Mr A. S. Wisdom

Can I just deal with one aspect of what the last speaker was talking about, namely this implied ignorance of Common Law rights, even to the suggestion that this island is full of ignorant people who have never seen a lawyer and do not even know they exist. The point is that people who wish to exercise rights, riparian rights, are usually owners of property, and there are, of course, national bodies such as the National Farmers' Union, the Country Landowners' Association, the Anglers' Co-operative Society, and so on, who are, I will not say stuffed with lawyers, but at least have access to legal facilities. They know when to exercise them, and I am not certain they do not, at times, over-exercise their Common Law rights, to my own knowledge.

Miss M. M. Sibthorp

Can I make just two short points, I am afraid unsupported at this moment, but I can support them. Quite recently a statement was made by one of the authorities that, although the air was cleaner, it contained more sulphur dioxide, which was increasing rather than decreasing. The second point is that whereas the foaming on rivers had, in fact, very much improved, recently it was again on the increase. May I also ask one question? What is the effect of the new biological ingredient in the detergent which has been recently produced which is supposed to destroy stains? I heard that it was quite deleterious as far as young children are concerned – it has an effect on the lungs – and I am wondering how far it persists in water.

Dr J. A. G. Taylor

The problem reported relating to the effects of enzymes on lungs referred only to workers on plants making these materials when these were first introduced and I think it should be stressed there has never been a danger to consumers. As a result of these teething troubles, rigorous precautions have been taken to ensure that safe working conditions exist. As far as the consumer is concerned, I think you can take it that there is no difference in your potential hazard between the two leading detergents.

As to the factor of longevity; once they go down the drain they meet conditions which only allow them to survive very briefly. This has been confirmed by work

135

in the USA. In any case there are many magnitudes, higher concentrations of proteolytic enzymes are to be found in normal sewage plants. Therefore, you can see, this is a non-problem.

DR J. R. GOLDSMITH
I would like to follow up Sir Frederick's question about what is the best way to obtain regulation, or evaluation or safe use of new chemicals, by suggesting that there are two examples that have been introduced into the discussion only this morning, for which some evaluation of hazards is needed. One that has been introduced just now has to do with enzymes in detergents. Now, and I am speaking on behalf of needed medical research – if perhaps I plead too strongly, I apologise – quite recently in Sweden it has been shown that there is an inherited blood enzyme deficiency, called the alpha-anti-trypsin deficiency. When it is in severe form it occurs in about only perhaps a few people out of ten thousand and leads to a chronic respiratory disease if individuals are exposed to respiratory irritants. So far, the case has only been proven for cigarette smoking, but suspicion, of course, attaches to community air pollution and, possibly, occupation exposure as well; occupational exposure to enzyme inhalation has been reported to lead to respiratory disease. So, it is conceivable that disease may arise in a small proportion of the general population who are unusually sensitive to the possible inhalation of enzymes of this sort. Now, I frankly would like to know, from those people here, how evaluation of the community implications of such occupational exposure ought to be carried out. Should it be done by mandatory provisions for reporting occupational experience, or should it be done on a voluntary basis? I certainly want to underscore the fact that, as far as I know, there is only evidence of some problems in the manufacture of this substance in the States. I do not know of any problems in its use. These occupational problems certainly deserve to be very rapidly investigated, and there should be an extremely full disclosure of what the facts are about whether exposed workers had this enzyme defect. This is not always done with new products, unfortunately. The employees who are engaged in the manufacture of a new product and who may or may not have health hazards as a result of this occupational exposure, obviously are people whose health reactions are of great importance to the general public. It is not always the case that this information is made available to the public health authorities, or to the general public.

The second example is the question of the nitroso-triacetic acid which has been proposed, I presume for very widespread use, as a detergent. I think that anyone who is concerned with nitrate pollution of underground waters and does not really know how the nitrate gets there, but is very much concerned about its health consequences, that such a person would rather like to know whether the use of such a compound widely in detergents is likely to increase pollution of underground waters with nitrates or substances which, on degradation, can yield nitrate. And there again, it seems to me the question is: will this be pursued in a mandatory fashion? Or are there adequate voluntary mechanisms which can ensure public health protection?

DR J. A. G. TAYLOR (Unilever Research)
I think, perhaps, I would like to come back on this NTA question which, I think,

is central again to us. NTA does degrade to nitrate. The change in the nitrate level one would expect by the use of this material is somewhere between 1 and 5 per cent, depending on the situation, and I think, again, we might look at the nitrate concern in this country. The present World Health Organization recommended standards are expressed in terms of nitrate-nitrogen, which is 10 parts per million. If you express this as nitrate, it is 45 parts per million. But they are currently, I understand, in the process of doubling this factor in a temperate climate such as ours, so this would make it very much higher. Now, we are going to be putting in, if this material were to come into use, I say if, up to an extra 5 per cent on the present situation.

On the question of the biodegradability of NTA there has been a good deal of work done by a number of companies. As Mr Owens will know, we are currently working with the Water Pollution Research Laboratory on a joint programme to clarify some of the outstanding issues. A good deal of work has been done in the United States, where the Procter and Gamble Company and others have also examined NTA in detail, so you can see that it is an area which has received considerable and detailed attention. The decision to use this material to replace phosphate is one that in the United States, in particular, involves social and political pressures more than technical ones, as the view that phosphates are the chief cause of algal blooms is still moot. If I can correct Mr Owen's figures, when he talked of the percentage of phosphate arising from detergents he referred to effluents. He should instead have spoken of domestic sewage effluents. Whilst up to 50 per cent of phosphates in the sewage effluents does come from this source, it should be borne in mind that only 13 per cent of the phosphate produced in the United States go into detergents.

Whilst a lot of the phosphate that is spread on to the land is retained, there must also be a contribution to the overall load. In fact the minimum requirement of phosphate for algal blooms is so small that I understand, from evidence given at a recent conference, that if you were to take a stream or river and literally block off both ends, clean it out completely and then put in distilled water, then you would leach out sufficient phosphate from the underlying ground to provide for algal growth.

Now the situation is this: sufficient phosphate is going into the environment from the human digestion to cause a many-fold excess over the minimum phosphate requirement for algal growth and the extra that detergent adds, I suggest, can do very little to alter the position. It is a material, as Mr Owen agreed, that we know a good deal about from long usage and hence has a lot to recommend it.

Dr D. W. Bowett, Chairman
In terms of the evaluation of the risk of pollution, I take it that where Unilever carries out an evaluation, it does so on a purely voluntary basis?

Dr J. A. G. Taylor
Yes, this Standing Technical Committee on Detergents operates on a voluntary basis in this country, but we have to take into account certain other situations in other countries. For instance, the soft detergent story in this country is one of agreement; in Germany and Spain there are laws that the materials should be

at least 80 per cent biodegradable, and I think this standard is probably going to be adopted by OECD countries. In fact most detergents are 90 to 95 per cent biodegradable.

Dr D. W. Bowett, Chairman
Well, that was really the question that Sir Frederick posed; whether it ought to be mandatory by statute, and, of course, it brings us back to your question as to whether this is really the way to proceed.

Sir Frederick Warner
Well, I think this is the whole question, if I may interrupt here. Ten parts per million *is* mandatory. By the introduction of this new material, it is pretty certain that it will have to go up to 20 parts per million, or this will jeopardise the use of the new material under certain circumstances, and I think the point raised is terribly important. We *have* a level now that is agreeable, but if pressure is brought on us to double that level – that you lawyers have agreed to put into law – then this opens the way for the new material, but possibly the way to more pollution.

Dr J. A. G. Taylor
I think that my remarks were misunderstood. I was not advocating that nitrate levels should be doubled but that the WHO had recommended for other reasons that this should be done in temperate countries. If NTA were to be used then the increase in nitrate level would only be up to 5 per cent of the current level, i.e. up to 0·5 ppm of nitrate nitrogen.

Sir Frederick Warner
Nobody foresaw that as nitrates were introduced into agriculture as NPK and so on, we would be faced with the nitrate position that has been brought up at international level. This is the real point, and we have to decide now, do we hang on to the mandatory levels that have already been agreed and established – this is your point, I should imagine – or do we find an alternative method? This is a very important thing.

Mr J. F. Whitfield
I would like to add a small footnote on dieldrin. My Committee came up against this in connection with local carpet manufacturers who, as you may know, use this substance as an anti-moth agent. On the Continent a less potent chemical is used and I believe the use of dieldrin is prohibited.

I am not a chemist and I have forgotten its name. Possibly someone can tell me and also what results have been obtained from tests in this country.

Mr M. Owens
As I understand it, carpet manufacturers in this country are considering switching from dieldrin to eulan for mothproofing, and in fact a number have already done so. The eulan utilised is a sulphanilide compound which is claimed to be biodegradable and some attempts have been and are being made to test this in sewage treatment. Some effluents containing eulan are discharged after treatment into the River Stour and thence into the River Severn, and the

WPRL is hoping to look at the effect of further changeover from dieldrin to eulan on the present levels of dieldrin residues in these rivers.

Dr C. G. Dobbs

I should like to get back to the general question of whether voluntary control will work. In Common Law – the sort of instance which Professor Garner used was that of the individual against Joe Bloggs downstream – that sort of thing – whereas in the modern world major pollutions are likely to come from the big industries. Any action here, as Professor Garner said, will depend on the willingness of the offended party to become the plaintiff, which, of course, he is unwilling to do against a powerful concern, be it a government department or a major industry or even the two acting together. We know that public enquiries do provide, or are supposed to provide, some sort of provision for this situation; and we also know that most public enquiries are rather a farce, because the top legal talent and the heavyweight scientific consultants are all on one side against only the local talent on the other. Stanstead is a good example, of course, where they happened to have a lot of influential local people with knighthoods and what not rallying round. If you do not have that you get nowhere. Now what I suggest is, simply, that at public enquiries there should be provision for legal aid, and, if necessary, scientific consultancy, to put the case against pollution – to put the case for the conservation of that particular locality, of equal weight with that against which they are arrayed. After all, individuals, even criminals, can get legal aid – but here scientific aid may be the more necessary.

This might make a voluntary system workable, and the present existing system work better. Furthermore, nearly all the examples that we have heard about, cases where there is any sort of public alarm and action, have been catastrophic. If people die from smog or mercury poisoning, if infants turn blue and die or are otherwise slaughtered, people actually notice! It appears to be perceptible, apparently, even to our Public Health Authorities! But anything as extreme as this is not usually what people are afraid of when, for example, a large smelter is being put near them, or something of that sort. Effects can be at what one might call the *predominant factor level* at which the injury is blatant and you can prove your point – and the law takes account of this. But this is unusual. What people are afraid of, mainly, are effects at the *contributory factor level*, where the effects of the pollutant are so mixed with those of other factors that you cannot prove your point in law, though you may strongly suspect that you are suffering injury. This is the major trouble; the catastrophic effects are of quite minor importance, in total, I suggest, in relation to this; and furthermore, synergistic or additive effects must be presumed to occur between different injurious factors even though they have been investigated and proved only in relatively few cases, so far. The result of this sort of situation – here we must bring in the idea of normality again – is that you always make a comparison with "normality", and provided the increase in harm is not such as to produce effects at the *predominant factor level* which you can prove, then it is assumed that no effects have been produced. Your idea of "normality" next time is adjusted imperceptibly, it is adjusted to the new situation because "normality" has covered the effect, and so we get an escalation of "normality" towards higher and higher pollution.

Now all these points want putting in a more authoritative way than I can do, and on much more frequent occasions. In conclusion may I repeat the suggestion that, at public enquiries, if they are not to be a complete farce in regions where there doesn't happen to be a lot of local talent to face the international talent which is liable to be dragged in from overseas and elsewhere, there should be some sort of provision for ensuring that the case against pollution is presented at least on equal terms with the case for it.

MR M. OWENS

Perhaps I might comment on the national and international organisation of research effort on pollution with particular reference to the work on enrichment or eutrophication. Nationally, the Fresh-water Sub-committee of the Natural Environment Research Council has set up a working group to review the extent of this problem in this country, to make recommendations about the research which is required, and to advise on the financing of that research and also on the extent of the effort required. Internationally, the work on eutrophication has been organised under the auspices of OECD. Their Water Management Research Group commissioned a consultant to review the whole field with particular emphasis on nitrogen and phosphorus. Delegates from European countries, Canada and the USA have met and discussed this review and have suggested amendments to it. The same delegates have drawn up a programme of what they considered to be the priorities which should be tackled and these, together with the report, were submitted to the Water Management Research Group. The WMR Group has recently decided to establish four working parties to advise on suitable research programmes; these are concerned with: (i) analytical and survey methods; (ii) methods of nutrient removal; (iii) detergents; (iv) fertilisers and agricultural drainage. It is intended that these working groups will attempt to harness international research and knowledge to alleviate the problems arising from enrichment of surface waters.

The David Davies
Memorial Institute of
International Studies

The Department of International
Politics of the University College
of Wales, Aberystwyth

CONFERENCE ON

LAW, SCIENCE AND POLITICS:
WATER POLLUTION AND ITS EFFECTS
CONSIDERED AS A WORLD PROBLEM

Sunday 12 July
AFTERNOON SESSION

Pollution by Industrial Waste and Sewage

Chairman (i) SIR FREDERICK WARNER
Cremer & Warner, Engineers

(ii) LORD SIMON
Port of London Authority

Scientist MR A. J. O'SULLIVAN
Lancashire and Western Sea Fisheries
Joint Committee

Administrator MR R. F. PEARSON
Greater London Council

Lawyer MR IAN DRUMMOND
Trent River Authority

Pollution by Industrial Waste and Sewage: Scientific Aspects and some Problems of Control in the Marine Environment

Mr A. J. O'Sullivan (Lancashire & Western Joint Sea Fisheries Committee)

1. Introduction and Historical Background

Pollution by sewage, i.e. domestic waste, is a problem that has been with man a long time. Industrial waste is on the other hand more recent; it gives rise to problems different in some ways from those resulting from the disposal of domestic sewage, but as we shall see later, the effects can be closely linked.

Disposal of domestic sewage did not cause great problems while populations remained scattered or reasonably small. Pollution was very local, and little of the marine environment was affected apart from estuaries within the boundaries of or close to large cities such as London. It was only with the growth of large conurbations that domestic sewage disposal became an international, or even a national problem.

Industrial waste, however, because of the greater toxicity of some of the materials produced, rose to international problem status more rapidly, but still only in very confined areas of the seas. Two other factors accentuate the difficulties:

(i) some industrial waste is virtually non-biodegradable;

(ii) there is an increasing tendency to site industries on the coast for reasons of effluent disposal; thus leading to more intensified discharges of effluent. An additional tendency is for effluents to be taken to sea by ships and discharged outside the territorial limits, thus making the problem fairly and squarely an international one.

As a scientist on this panel, I feel that my first task must be to say something about the effects of sewage and industrial waste discharge, for only on the basis of this knowledge can we effectively discuss the problems of control. I hope later to mention briefly a few control and administration problems, without, I trust, treading on Mr Pearson's ground. The Sea Fisheries Committee for which I work is essentially a controlling and administrative authority; it has statutory powers under the Sea Fisheries Regulation Act of 1966, which superseded an earlier Act of 1888. These powers allow the Committee to prohibit or regulate the deposit or discharge of substances detrimental to sea fish or sea fishing.

2. Composition of Domestic and Industrial Waste

The effects of domestic sewage on the environment are determined to some extent by the contents of the sewage. Organic material is the chief constituent but the mixture also contains nutrient salts, silt and other inert suspended solids, pathogenic bacteria and viruses.

The passage of the Drainage of Trade Premises Act in 1937 gave industry the right under certain conditions to discharge sewage into public sewers and this has resulted in most town sewage containing industrial waste. A recent calculation shows that on average about half the volume of sewage received at local authority sewage works in this country is composed of industrial effluent. In this way some of the materials already present in domestic waste are added in further quantities and, in addition, trade waste may contain conservative materials, that is, substances that are non-biodegradable or are broken down by bacteria only very slowly. These materials include heavy metals, organic chemicals, pesticides and synthetic detergents.

So when we refer to domestic waste we have to consider a mixture, each constituent of which has different environmental effects. Industrial waste is also usually a mixture of process effluents; in some cases so much so that chemical analysis is very difficult and biological testing of the toxicity of the mixture yields more valuable results.

3. Specific Effects

3.1. *Organic Matter*

Pollution by organic matter includes effects due to oxidisable materials, nutrients derived from their breakdown and a significant amount of suspended solids. The immediate effect of oxidisable materials is to reduce the amount of oxygen in the water; both by providing an energy source for micro-organisms which use oxygen to respire, and by simple chemical action. Many species of animals are eliminated, while those that remain may become very numerous. This is particularly true in sheltered and semi-sheltered areas of the sea and estuaries where extremely high numbers of minute oligochaetes have been recorded in polluted conditions. Under conditions of gross pollution, however, an area becomes completely anaerobic and devoid of all life above the level of bacteria. Under normal conditions domestic sewage is rapidly oxidised and outside the immediate zone of discharge an area rich in animals is found (Tulkki 1968; Leppakoski 1968). Conditions surrounding the area of discharge can be described as a result of two factors:

(i) The rate at which material is added, and
(ii) The rate at which it is dispersed, diluted and oxidised.

In the open sea these latter factors are very great and it is rare that the oxidative capacity of the receiving water is exceeded. Nevertheless there are a number of sea areas where conditions have become very bad. Pearce (1969), in a preliminary report of investigations into the effects of dumping sewage sludge off New York, found that an area of between 31 and 47 sq. kilometres is almost devoid of animal life with a very low standing crop of a few species of organisms. Other areas in which addition is beginning to exceed oxidative capacity and which are the subject of increasing concern are the Great Lakes in North America,

144

the Caspian Sea, the Baltic Sea, the Sound between Sweden and Denmark and more recently the North Sea.

3.2. *Nutrient Salts*

Even where full biological treatment is given to domestic sewage, such treatment does little to remove nitrates or phosphates from the effluent. It has been estimated (Painter and Viney, 1969) that the amount of phosphorus and nitrogen contributed by each person every day is 2 grams and 9 grams respectively. Hynes (1960) calculated that each day the population of Great Britain wastes approximately 500 tons of nitrogen and 30 tons of phosphorus in sewage effluents alone. In the Lancashire and Western Sea Fisheries District, the coastline of which is 440 miles long, a survey carried out in 1968 estimated the amount of nutrients entering the sea from local authority sewers as 1,068 kg of nitrogen per day and 2,240 kg per day of phosphorus in winter, and 15,188 kg per day of nitrogen and 3,275 kg per day of phosphorus in summer. These figures do not include trade wastes nor the amount of pollutants entering the sea from the discharge of rivers. About 21 per cent of the phosphorus in sewage is contributed by synthetic detergents.

Nutrient salts are the cause of the problem of eutrophication or over-enrichment. When excess nutrients are present the growth of microscopic green algae is no longer held in check and "plankton blooms" can be the result. Minute flagellates are encouraged and some species of these liberate toxins which kill or stun fish. If shellfish ingest these flagellates they too harbour the toxin and can be the cause of paralytic shellfish poisoning if eaten by man.

Even when the bloom is caused by non-toxic species of algae the amount of organic material involved is quite large and can give rise to anaerobic or toxic conditions, particularly if senescence and decay of the bloom occur in shallow water. This happens quite frequently on the north coast of Wales and the Sea Fisheries Committee for whom I work immediately receives reports of industrial effluent or sewage pollution on the beaches. In this case I am happy to inform the industrialists that it is not their fault!

Recently (June 1970) a dense bloom of *Phaeocystis* gave rise to the un-precedented phenomenon of large amounts of foam blowing ashore at Blackpool. Newspapers described the Promenade as being a foot deep in foam in places and worried corporation officials rang the Sea Fisheries Committee fearing, as they thought, that it was the result of spraying a massive oil slick at sea. In fact the occurrence was the result of extremely warm conditions which allowed large quantities of *Phaeocystis* to proliferate in the top few centimetres of the sea. With the onset of a sudden gale, the mass of plankton was driven ashore before it could disperse and, being churned up by the heavy surf, led to the formation of large amounts of foam.

The problems that therefore arise as a result of wastes containing nutrient salts being discharged into estuaries and inshore marine areas come not from nutrients themselves but from the changes they induce in the productivity and in particular the phytoplankton of the area. Eutrophication of lakes by sewage and land run-off has been the subject of much research and concern in recent years but despite the fact that all nutrients end up in the sea, much less is known about the

145

essentially similar problems that occur in the marine environment. It is true to say that the problems are not as severe and are the cause of noticeable or detrimental effects only in semi-enclosed areas such as estuaries, fjords and certain bays. Oslo Fjord in particular has been seriously polluted by sewage effluent for many years to the extent that fisheries have declined and shellfish have accumulated toxins as a result of dinoflagellate blooms. Off the north-east coast of England the outbreak of paralytic shellfish poisoning which occurred in the months of May and June 1968 as a result of a bloom of dinoflagellates could very well have had its primary cause in the excessive amounts of pollutants discharged to the area. There is no direct evidence for this except perhaps a purely associative sort in that plankton blooms seem to be confined to inshore areas close to major polluting sources. According to Clark (1968) the reasons for occasional population explosions of planktonic organisms are not fully understood but depend on the existence of rich supplies of inorganic nutrients. Korringa (1968) lists several occurrences of such poisoning and points out that in nearly all cases it has been connected with eutrophic conditions.

In the United States, Boston Harbour and the Potomac Estuary have shown signs of eutrophications; in the former area the excess growth of the seaweed *Ulva lactuca* has become a considerable problem and has been ascribed by Sawyer (1965) to nutrients derived from domestic waste. In Chesapeake Bay *Ulva lactuca* also appeared to be excessively prevalent in areas receiving domestic wastes and caused secondary pollution due to its becoming detached and concentrated by wind and tide. The decomposition of the weed caused the water to become opaque and creamy in colour. Hydrogen sulphide gas was liberated and there were reports of fish mortality. It has also been noticed from recent reports in the *Irish Times* that similar conditions appear to be on the increase in Dublin Bay. An incident reported on 20 May described hundreds of dead crabs and ragworms on a short stretch of estuarine shore and a letter to the same paper a few days later spoke of "acres of sea-lettuce growing on the north side of the bay". The letter went on to say that the weed was being torn from its mooring by the tide and was piling up on the shore, and its decomposition was producing hydrogen sulphide and other malodorous gases. It may be that before the end of the century the eutrophic conditions extending outward from Liverpool Bay and Dublin Bay will meet, and that the Irish Sea will follow the course of slow deterioration as is now happening in the Baltic.

3.3. *Silt*

Pollution by inert sewage-derived solids and by the particulate trade wastes can give rise to undesirable effects, particularly in estuaries and nearshore coastal waters. Very little work appears to have been done on this aspect of waste disposal ecology compared with the amount done on the diffusion and dispersion of the soluble portions of the waste. The behaviour of suspended solids can be quite different from that of dissolved materials and instead of being diluted and transported away from the coast they are frequently trapped and accumulated in the nearshore environment. Salinity changes and the presence of organic matter encourage the settling out not only of inorganic solids in suspension in sewage but also of other naturally occurring silt fractions. Two principal effects of silt are:

146

(i) smothering of benthic animals, and

(ii) the absorption of light and reduction of photosynthesis by phytoplankton.

In general benthic animals are tolerant of a wide range of silty conditions, especially those animals living in estuaries. Nevertheless any change in the rate of silt deposition or in its composition may have considerable ecological effects. A study of the effects of china clay waste in St Austell and Mevagissey Bays by Howell and Shelton (1970) indicated conditions of near sterility around the places of discharge close inshore. A little farther offshore, however, where the rate of deposition is presumably less, a rich community of deposit feeding animals has become established.

3.4. *Bacteria*

Bacteria in sewage constitute a potential hazard to human health. Direct contamination of bathers by sewage-derived pathogens is not supposed to be a very great risk according to the Medical Research Council (1959). Despite continuous investigation of this aspect of pollution and the accumulation of both bacteriological and epidemiological evidence there seems little reason to change that view. However, any increase in the total volume of sewage discharged into the sea will undoubtedly increase the risk.

Eating contaminated shellfish such as mussels which have filtered bacteria out of the water presents a much greater risk. Among the filter feeding molluscs which concentrate on bacteria are oysters, mussels, clams, cockles and scallops, and there have been many cases of enteric disease resulting from the consumption of these shellfish taken from polluted waters. To counteract this hazard to public health many local authorities have prohibited, under the Public Health (Shellfish) Regulations, 1934, the taking of shellfish from polluted areas unless the shellfish have been cleansed or sterilised. Under these regulations many shellfish grounds in estuaries or coastal waters have been partially or completely closed. In the Lancashire and Western Sea Fisheries district alone, fourteen closing orders or shellfish regulations apply. All of the shellfish beds close to populated areas from the Duddon estuary to the Menai Straits are subject to orders either of complete prohibition or subject to cleansing or sterilisation. Sherwood (1947) comments that "what were once flourishing public beds open to unrestricted fishing have ceased to exist".

No one denies the right of a public health official to protect the public from what he considers to be a health hazard, but it does seem anomalous that the fishing industry should have to bear either the cost of cleansing or the loss of what were once economically viable fisheries. If we look upon our seas and coastal waters as a natural resource it does not seem right for cities, towns and in some cases industries to be able to destroy part of that natural resource as far as one section of the population is concerned. This perhaps is why those who deal with sewage pollution envy the state of control in the field of radioactive pollution where a small number of citizens liable to be affected, as Mr Preston described, are the subject of special provisions. There are therefore sound ecological and humanitarian arguments, and a precedent, for the view that the authority responsible for pollution of the shellfish beds should undertake the

147

cost of remedial measures, i.e., should provide cleansing arrangements for the shellfish or else discharge its sewage in such a way that no further pollution is caused.

3.5. *Conservative and Non-biodegradable Materials*

Conservative and non-biodegradable materials constitute the greatest potential risk to the environment. The number of these materials particularly in the field, of biocides, polymers and chlorinated cyclic carbon compounds, of which DDT was one of the first examples, is steadily increasing. The danger is that they remain in the environment for a very long time and may become concentrated by physical or biological means so as to reach toxic levels in some marine animals and man. As in the case of the finding of polychlorinated biphenyls in seabirds in the Irish Sea during 1969 and 1970, they turn up in unexpected places. It is true that the necessary knowledge is in existence to suggest or show that this bio-accumulation could occur, nevertheless the findings were a surprise to many ornithologists and environmental scientists. Furthermore it was only after the discovery of polychlorinated biphenyls in predatory birds that a serious attempt was made to discover their origin and bioaccumulative pathways. Moreover, in recent years increasing amounts of metals discharged by industry have led to a number of serious pollution incidents. Mercury has been one of the worst offenders, being the cause of Minamata disease in Japan and also the subject of much concern in Sweden where its extensive use as a fungicide in the paper industry has led to the pollution of a number of rivers, lakes and harbours. I understand that there is still a harbour in Sweden, the bottom mud of which is so polluted by mercury that the harbour cannot be dredged because of the toxicity of the dredged material. In the United States mercury pollution was thought to be a problem only in the Great Lakes region, but it has since become a dangerous contaminant in thirty-three states. As in Sweden, the worst offenders are paper companies, or chemical companies using mercury cells to separate chlorine from brine solutions. In this country much of the mercury present in soil or water is of natural origin; there is not yet a serious problem, though it has been stated that a greater quantity of mercury per hectare is used for agricultural purposes in Britain than in Sweden.

So much has been written on the way in which pesticides have become part of the global ecosystem that I will not go into here, save to mention that DDT, which is handled so carelessly and in such large amounts, is extremely toxic to marine organisms. Manwell and Baker (1967) cite a paper by Grosch in which it was reported that *Artemia salina* is killed by DDT at the fantastically low level of one part DDT in 100,000 million (1×10^{11}) parts of salt water.

4. Control and Administration

Control of water pollution is an extensive subject but in this brief paper I feel I ought to limit my remarks to conditions applying in the marine environment which is the particular aspect with which I am most concerned. There is as yet in this country very little effective control over pollution of the sea. The Salmon and Freshwater Fisheries Acts give river authorities some jurisdiction over waters, including marine waters, containing salmonid fish but the existence of these Acts are of little value as they relate only to direct damage to salmonid fish and take no account of environmental degradation. The statutory powers of local Sea Fisheries Committees are of greater value in that they can prohibit or regulate the deposit of substances detrimental to sea fish or sea fishing. Interpretation of this power varies among the eleven sea fisheries committees around England and Wales but a number of them take what could be considered an ecological viewpoint, and regard substances that affect, say, the larval biology, incidence of disease, availability of food or habitats to sea fish to be detrimental substances. A new form of by-law which was evolved by the Association of Sea Fisheries Committees and which it is hoped will shortly be put into effect by a number of committees, codifies this ecological viewpoint and may be considered as one of the very first positive legal steps towards dealing with deterioration of the marine environment. The ultimate effect must still be in terms of detriment to sea fish or sea fishing activities but since fish may be regarded as lying at the end of an ecological cause and effect chain, most forms of marine pollution will have some affects upon them. The legislation still suffers from two disadvantages, however:

(i) control extends only as far seaward as the three mile limit of territorial waters, and

(ii) the by-laws do not apply to the discharge of domestic sewage by local authorities under their statutory powers.

The first of these two limitations is of particular importance in view of the increasing tendency for sewage waste, sludge and industrial effluent to be taken by ship to deposit areas outside territorial limits. It would be difficult for sea fisheries committees to maintain effective control over such dumping at sea; the only possible solution I could envisage would appear to be a form of control at the quayside or dock where the vessel was loaded. The difficulty of overseeing accurately where the material was being dumped could be mitigated to some extent by rules requiring the keeping of records. Such rules are already in existence under the Oil and Navigable Waters Acts governing the pumping of bilges, tank cleaning and transferring of oil. Although by no means foolproof, records, if properly examined, could be of considerable assistance in the control of dumping and a tight check on the duration of each voyage should remove most economic incentives for dumping closer to land than permitted. Also the activities of these effluent tankers are frequently reported upon by aircraft, mainly because the appearance of the effluent "slick" resembles oil when seen from the air.

5. Future Trends and Possibilities

The important fact that we are forced to accept is that practically all of man's efforts today are tending towards the simplification of the total ecosystem, that is, towards a reduction in the variety of habitats and species around us. For example, in agriculture the practice of monoculture is a drastic simplification of what is normally a very diverse system of plant life. Antibiotics and insecticides are also simplifications in that they are replacing, by indiscriminate killing agents, the

149

complex controls that normally keep insect populations in check. Pollution of the environment is one of the most important ways in which this natural diversity is being reduced. The flow of energy which can come initially only from sunlight follows complex pathways in natural systems; the introduction of organic pollutants short-circuits this energy flow. Energy is provided by man's wastes in the form of chemical bonds, e.g., sugars, proteins, hydrocarbons, and the natural ecosystem adapts from one fuelled by the energy of sunlight to that fuelled by the oxidation of organic waste. A system such as this favours microbes which in Nature may be pre-adapted to dealing with chemical pollutants. It is well known that complexity in ecosystems gives them more stability and thus the simplification brought about by man leads to both men and the ecosystem on which he depends becoming more vulnerable to environmental changes.

Short-term decisions taken many years ago are now leading to environmental problems. For example, the decision to allow trade waste to be sent to local sewage works, while a good decision from the point of view of the ease of treatment, has resulted in a large proportion of sewage sludge being unusable as fertiliser on the land. Sludge from works receiving only domestic sewage has some value but when it contains heavy metals and toxic chemicals derived from trade waste which are not broken down entirely in the sewage treatment process it becomes a slow poison. It may be that this earlier decision will have to be reconsidered or that some initial treatment of trade wastes should take place at source.

Some form of treatment for all sewage being discharged to the sea has been suggested but even if this was carried out, the problem of eutrophication would still be only partially reduced. At present about 37–46 per cent of phosphate is removed from conventional biological treatment; removal of higher percentages demands either an extra stage in the treatment process or the addition of salts of calcium, iron or aluminium during some stage of the conventional process. Both methods increase the cost of sewage treatment, yet it is understood that in Sweden all sewage works will be required to remove phosphate from their effluents by 1975. The additional treatment facilities necessary will be subsidised by the Swedish Government; the extent of the subsidy depending on the percentages of BOD and phosphate removed. It may be that such a scheme would be suitable for relieving the eutrophication problem in vulnerable areas in this country.

On the other hand it may prove possible to make use of the fertilising properties of sewage in some areas, with the proviso again that the discharge must be free from conservative materials. This use of sewage has been recognised as an aid to the culture of food-fish in freshwater ponds; it may soon be possible to find a similar use in the marine environment, particularly with the growth of fish farming. An intermediate culture organism on which the fish can feed may be necessary for metabolism of the waste, and dangers to guard against would be those resulting from bio-accumulative household chemicals and from transfer of parasites.

Reclamation and recycling of any kind is better than pollution. The technology is available, but the reason why pollution continues is that it appears cheaper to pollute than to reclaim! Thus economics has the final word in the argument, but the economic factor is itself a function of the relative costs of treatment and extraction of raw material. Thus the advent of cheaper treatment processes or a drying up of the supplies of raw material may swing the balance to reclamation.

Meanwhile, what can most immediately be done to put a halt to pollution, marine or otherwise? Firstly there is the greatest need for a rapid increase in the research effort being devoted to basic and applied problems in ecology, oceanography and in the dynamics of the ocean-atmosphere system. Here I hope that the forthcoming International Decade of Ocean Exploration will play an important part. Secondly, those who make decisions having effects on the environment must learn, or be taught, to foresee as many as possible of the consequences of such decisions. In other words we need a kind of ecological education for administrators, politicians, engineers and many others. Thirdly and lastly, we need the kind of interdisciplinary studies that bring together biologists, chemists, engineers and administrators when confronted with an ecological or pollution problem. In this way not only will decision-makers be ecologically educated but ecologists will be enlightened with regard to the processes of administration and law. This, it seems to me, is what this conference has aimed at and is now making a success of.

References

CLARK, R. B. (1968). Biological causes and effects of paralytic shellfish poisoning, *The Lancet*, 5 Oct. 1968, 770–2.

HOWELL, B. R., and SHELTON, R. G. J. (1970). The effect of china clay on the bottom fauna of St Austell and Mevagissey Bays, *J. Mar. biol. Ass. U.K.*, **50**, 593–607.

HYNES, H. B. N. (1960). *The Biology of Polluted Waters*, Liverpool University Press, 202 pp.

KORRINGA, P. (1968). Biological consequences of marine pollution with special reference to the North Sea fisheries, *Helgolander wiss. Meeresunters.*, **17**, 126–40.

LEPPAKOSKI, E. (1968). Some effects of pollution on the benthic environment of the Gullmarsfjord, *Helgolander wiss. Meeresunters.*, **17**, 291–301.

MANWELL, C., and BAKER, C. M. ANN (1967). The pesticide problem, *J. Devon Trust for Nat. Conserv.*, **12**, 491–3 and 498.

MEDICAL RESEARCH COUNCIL (1959). Sewage contamination of bathing beaches in England and Wales, *Memorandum No. 37*, London (HMSO), 24 pp.

PAINTER, H. A., and VINEY, M. (1959). Composition of a domestic sewage, *J. biochem. microbiol. Technol. and Engng*, **1**, 143–62.

SAWYER, C. N. (1965). The sea lettuce problem in Boston Harbour, *J. Wat. Pollut. Control Fed.*, **37**, 1122–33.

SHERWOOD, H. P. (1947). *Safe Shellfish*, Ministry of Agriculture, Fisheries and Food, London, 13 pp.

TULKKI, P. (1968). Effect of pollution on the benthos of Gothenburg, *Helgolander wiss. Meeresunters.*, **17**, 209–15.

PEARCE, J. B. (1969). Investigation of the effects of sewage sludge and acid wastes on offshore marine environments, *Marine Pollution Bulletin*, **7** (Jan. 1969), 5–7.

151

The Greater London Council's Work in the Control of Water Pollution

Mr R. F. Pearson (Greater London Council)

London provides an outstanding example of the development of sanitation with the growth and intensification of communal living. A study of the history of this development shows how it has been achieved through the combination of Law and Science processed by Politics. It shows how, as the individual could not control sanitary conditions, the matter had to be taken from the region of private initiative to that of public administration, and may serve as a guide for the further development of national and international co-operation in pollution control.

Historical

Until about 1830, town improvements of the previous century, humanitarian activities, advances in medical knowledge, an abundant food supply, rising urban wages and the work of Improvement Commissions had more than kept pace with sanitary problems arising from the concentration of population in towns of ever-growing size, and the death rate continued to fall until the 1820s and 1830s. Then, as the great towns continued to grow, the sanitary evils generated far surpassed the then existing capacity for improvement. Between 1831 and 1841 the average death rate per thousand of five of the larger manufacturing towns in England rose from 20·7 to 30·8.

Over the larger part of towns there was no provision, either municipal or private, for any means of cleansing at all. Water, whether obtained from a river, a water company or a well, had to be stored in kettles or buckets except in wealthy houses where tanks were used. Sewerage, in the modern sense, did not exist; such sewers as existed were for carrying off rain water and, there being no continuous water supply, filth, excrement and other decaying matter were normally removed only when rain washed them into the ditches and surface water culverts whence they eventually reached the streams and rivers. Cesspools were emptied at most three times a year, but usually once in every two years, by men with shovels or buckets and ropes, into carts which were then driven away and emptied into the rivers. Few houses had privies, and even those which had might have nowhere to deposit their contents. There were houses whose yards were completely covered with human ordure 6 in. deep, across which the inhabitants stepped on bricks.

It is not surprising that epidemics of "fever" grew in these conditions. London suffered severely in the cholera visitations of 1831/32, 1848/49 and 1853/54. In 1849 the deaths from this cause were 18,036 and in 1854 nearly 20,000.

"The Sanitary Report" of 1842, written by Edwin Chadwick, touched off public consciousness of the problem and with Chadwick's drive as the mainspring, a Royal Commission was set up in 1843 to enquire into the health of towns. The work of this Commission was the basis for the subsequent legislation of the 'forties and 'fifties.

Chadwick's solution of the problem was a system in which water supply, house drainage, street drainage and main sewerage were used to convey the waste matters from streets and houses, using water as the medium for transport. The system entailed carrying water into every house, and the removal of all excreta in suspension in water by means of the soil pan.

To provide for the practical implementation of his scheme Chadwick showed that, as the individual could have no control over conditions, the matter must be taken from the regime of private initiative to that of public administration. The outcome of Chadwick's work was the passing of the Public Health Act, 1848.

The crisis through water-borne pollution in London and some other large towns which led to the "health movement" of the 1830s may be regarded as a forerunner to the crisis towards which the world is now moving. The Public Health Act of 1848, in attempting, although by no means perfectly, to lay down a minimum of sanitary services was a landmark in public health history.

For London, as it was then constituted, the Metropolitan Management Acts of 1855 and 1858 provided the powers necessary for the systematic drainage of the county. In 1856 the Metropolitan Board of Works was appointed under the first of these Acts with the primary duty of maintaining the main sewers and constructing works to prevent sewage entering the Thames within the London area. The Board's work consisted essentially of the construction of "intercepting" sewers on both sides of the river and parallel to it, to intercept the old main line sewers in their courses to the Thames, and of "outfall" sewers which in turn discharged the sewage into the river below London at Beckton and Crossness. Following the recommendations of a Royal Commission set up in 1882, construction of plant for treatment of the sewage by sedimentation was commenced by the Board in 1887 and completed by its successor, the London County Council, in 1891. The system introduced into London the first regional scheme in the world for drainage and sewage disposal.

The Greater London Council was established under the London Government Act, 1963, which became effective on 1 April 1965.

Greater London extends approximately 36 miles east to west and 30 miles from north to south. Its administrative area is about 620 square miles and its population about 8 million at an average density of about 13,000 people to the square mile.

The 1963 Act provided for regional control of the sewerage and sewage disposal of the whole area and for the rationalisation of the service as it then existed. Rationalisation of sewerage and sewage disposal with a view to the concentration of sewage treatment at large, modern, economical works so that small obsolete works could be closed and valuable land released for other purposes, had been started before the formation of the Greater London Council. Notable examples of this are two schemes carried out by the Middlesex County Council which resulted in the replacement of twenty-six small works by the Mogden Works in 1936 and ten by the Deephams Works in 1963. The whole of the sewage from the 120 square miles of the old London County Council area was concentrated at Beckton and Crossness when the sewerage system for that area was first brought into being in the middle of the nineteenth century.

Under the London Government Act, seven of the twenty-one sewage disposal works which served the Greater London administrative area were transferred to the Council on 1 April 1965, and one, that of the West Kent Main Sewerage

153

Board, was excluded from the Council's jurisdiction. The Council's sewerage area then, mainly within Greater London but extending to some areas outside it, was about 500 square miles; it served a population of about 7 million and had a daily average sewage flow of about 450 million gallons.

With a view to rationalising and unifying the various drainage and disposal systems, the Act also required the Council to review the remaining sewerage and sewage disposal arrangements of Greater London, to consider which other sewers and sewage disposal works should be transferred to it, and to give notice by 1 April 1970 of its intention to vest them.

As a result of this review the first stage of the rationalisation of Greater London's sewerage and sewage disposal envisaged in the Act has now practically been completed. Fifteen of the twenty disposal works and their main sewers are vested in the Council, three have been closed and their flows diverted to the larger works, and the remaining two will be closed within the next two years.

The reorganisation makes the Council responsible for main sewerage and sewage disposal for an area of about 630 square miles; the population served is about 8 million and the daily average sewage flow about 550 million gallons.

In the next stage of the rationalisation it is proposed that further of the smaller works should be closed down until ultimately the Council's sewerage area is served by seven disposal works.

Drainage Systems

The Council's main-drainage responsibility in the Greater London sewerage area consists in the reception of foul sewage and, in some cases, surface water from the local sewers provided by other authorities into its main-line, trunk and outfall sewers; conveyance of the sewage, by gravity and pumping, to sewage treatment works; and disposal of the effluent and sludge produced at these works after treatment of the sewage.

Sewage, which is 99·9 per cent water, consists primarily of fluid domestic waste and industrial discharges; these are invariably connected to foul sewers. In many areas rain-water from roofs, roads, and other impermeable locations is also discharged into the foul sewers (the combined system); elsewhere storm-water is drained into separate surface-water sewers discharging to watercourses (the totally separate system); in other areas only part of the surface-water run-off enters the foul sewers (the partially separate system).

In combined systems the storm-water flow may on occasion be up to forty times the average daily flow of domestic and industrial wastes, and as it would be uneconomic to build sewers for flows of this magnitude the excess is usually discharged over weirs or by pumping to the nearest river or other watercourse. When this happens the sewage is, of course, heavily diluted with storm-water; moreover, the storm-water is discharged at such a rate as to cause the minimum practicable pollution of watercourses it enters.

Inner London

The sewerage in what was formerly the main drainage area of the London County Council is almost entirely on the combined system, and is characterised by long intercepting sewers running generally from west to east and flowing eventually to the sewage treatment works at Beckton (Newham) north of the River Thames and Crossness (Bexley) south of the river, which together deal with about two-thirds of the sewage flow from the main drainage area transferred to the Council in 1965. From these two works the treated effluent is released into the River Thames and the remaining sludge is taken by specially designed vessels for discharge into the North Sea. The capacity of the intercepting sewers, most of which were built between about 1859 and 1908, was generous; for the most part they have been able to carry the constantly increasing sewage and storm-water flows caused by the rapid development of London and an increased rate of water consumption, but overflow reliefs to the rivers and additional storm-pumping capacity have had to be provided from time to time.

The capacity of the inner London system is now below the accepted standard of six times the dry-weather flow, and new intercepting sewers are needed to reduce pollution loads and improve unsatisfactory conditions in some of the older sewers. New sewers are also needed to deal with the continuing intensive development of areas such as marshes, waste land, railway land and docks that previously made only small demands on the main drainage system. The present limitations on capital expenditure have, however, entailed some postponement of schemes desirable to maintain efficiency and to reduce river pollution and operating costs.

Outer London

Those parts of the sewerage area transferred in 1965 which are outside inner London, drain to the following sewage treatment works: Deephams (Enfield) and Mogden (Hounslow), Hogsmill Valley (Kingston upon Thames), Kew (Richmond upon Thames) and Wandle Valley (Merton). Since 1965 the following treatment works have also become vested in the Council: Beddington (Croydon), Redbridge Eastern and Redbridge Southern (Redbridge), Bury Farm (Havering), Sewardstonebury (Waltham Holy Cross), Sutton and Worcester Park (Sutton), Riverside (Barking). Here the drainage systems are separate or partially separate, and the surface-water sewers are generally vested in the local borough.

In outer London the maximum flow that local authorities may discharge to the Council's sewers is subject to statutory limitation only in the areas draining to Mogden and Deephams, but this maximum is often exceeded and before long it will be necessary either to insist on more rigorous observance of the limits or to provide more sewerage capacity. Elsewhere it was generally the case before 1965 that the local authority either owned the sewage treatment works or was represented on its board of management, and the authority thus had a direct interest in so arranging its drainage system as to avoid the discharge of excessive storm flows to the sewage disposal works. This is not now the case, and early action is needed, by legislation or otherwise, to ensure against progressive abandonment of the separate system of drainage in these areas.

Sewerage and Pumping

In 1965 the Council became responsible for the operation and maintenance of about 530 miles of main sewers and eleven large pumping stations. A further

155

140 miles of main sewers have since been vested. On conclusion of vesting there will be some 700 miles of sewers and fifteen major pumping stations. In addition, there are sewage pumping stations at most of the disposal works.

Some of the major sewers, particularly in inner London, have been in use for over 100 years and most for more than fifty. Because of their age and methods of construction, they need constant maintenance and improvement to bring them up to the required operational, structural and safety standards; in many of them, moreover, slack gradients and overloading necessitate regular cleansing. The sewers in the Middlesex area were mostly laid in the middle 1930s and in general present fewer problems of maintenance. Preliminary investigation of those taken over from authorities in Surrey shows that much work will be needed to bring them up to the required standard and to reduce overloading and the risk of flooding.

Pumping stations may perform two functions: to lift sewage from low-level to high-level sewers, thus allowing the flow to gravitate towards the treatment works; and to discharge excess flows direct to the river. Some stations serve only the second purpose. The oldest stations date back to the 1850s, and for some a programme of modernisation or replacement is in hand.

The Council has carried out a number of projects to improve the sewerage of its area and reduce flooding, notably the following:

New pumping station and sewers in Hammersmith (£2½ million).
New relief sewer in Greenwich (£1 million).
Reconstruction of sewers in the Isle of Dogs (£1 million).
Diversion of sewer from Ham area of Richmond upon Thames (£0·3 million).

Other major schemes on the drawing board include:

Diversion of sewage flow from Redbridge Eastern sewage treatment works to Beckton works (£3½ million). This scheme is in accordance with the Council's policy of concentrating sewage treatment in relatively few of the larger works, but its timing has to be advanced to allow for the construction of the proposed M11 motorway, the route of which lies across the Redbridge Eastern works.
New trunk (deep intercepting) sewers north and south of the Thames to reduce river pollution and the risk of flooding (£30 million).
Replacement of the obsolescent Abbey Mills pumping station and the construction of storm-water tanks (£7 million).

Due to financial considerations, construction work on the two last-named schemes is not expected to start for some years.

Sewage Treatment and Disposal

About one part in a thousand of sewage consists of suspended and dissolved solids, partly mineral and partly organic. Treatment consists, in the first instance, of separating the sewage into two constituents, effluent and sludge, that are dealt with by various processes before disposal. The method is, in principle, much the same at all works:

(i) Screening, to remove the larger solids: these may be cut up by disintegrators and returned to the flow, or instead of being screened they may be cut up by comminutors without being removed from the flow;

(ii) Removal of grit (or detritus) – passage of the sewage through channels at a controlled velocity causes settlement of the heaviest inorganic material, which is then dredged out, washed, and disposed of inoffensively on land;

(iii) Primary sedimentation in tanks, generally for four to six hours, during which about two-thirds of the remaining, more finely divided, suspended solids settle out and are then drawn off as sludge;

(iv) Secondary treatment (partly biological, partly chemical and partly mechanical) in aeration tanks. This transforms the dissolved impurities and suspended organic particles into "flocs" of activated sludge which can be readily settled out, leaving an effluent which is sufficiently pure to run off into a watercourse; and

(v) Sludge digestion, in which the sludge, generally containing about 95 per cent water and 5 per cent solid matter, is heated to about 85°F. in covered tanks. This, by a process of natural anaerobic digestion, converts the greater part of the organic solids into sludge gas (methane and carbon dioxide) and soluble matter, and reduces the total solids by about one-third. The sludge, moreover, loses most of its offensive properties, and much of the liquid constituent can be separated out. The remaining sludge is disposed of in different ways, according to the location of the works and the means available.

Bearing in mind that the objective is to carry out sewage treatment and disposal with the minimum pollution of rivers consistent with the cost the community is prepared to pay, the Council's efforts have been directed to continuing and improving the operation of the available plants, augmenting and modernising plants to give improved effluents and replacing equipment that is worn out or obsolete.

As mentioned later, the Council exercises pollution control over certain of the rivers into which the effluent from its sewage works discharges. The effluent from four of the Council's works passes into such rivers, while rivers under the control of other authorities receive the effluent from the remaining eleven works, including five which discharge direct to the tidal Thames.

Since 1965 the following major schemes have been, or are being, carried out at the Council's treatment works:

Beckton works (Newham) – modernisation and extensions to provide full treatment for the whole of the average sewage flow of 220 rising to 250 million gallons a day (£21 million).

Crossness works (Bexley) – increase in power for activated sludge plant (£$\frac{1}{2}$ million).

Deephams works (Enfield) – completion of original works as designed and extensions to cope with additional sewage flow and to improve quality of the effluent discharged to the River Lee (£5·3 million).

Deephams works – construction of pipeline through which sludge can be pumped to Beckton works for disposal at sea (£1 million).

Mogden works (Hounslow) – additional sludge digestion plant (£0·3 million).

157

Sludge Disposal

Sludge from the works at Beckton and Crossness is disposed of at sea by the Council's fleet of sludge vessels; that from Riverside sewage treatment works (Barking) is pumped to the Beckton works for disposal in the same way, and there will be a similar arrangement for the sludge produced at Deephams works on completion of the new sludge main connecting Deephams to Beckton. At the Council's other works the sludge is disposed of to land after air drying or mechanical pressing.

These methods are becoming unsatisfactory as the quantity of sludge increases. Air drying is subject to the weather and requires larger areas of land than are now readily available; moreover, the dumping of large quantities of sludge in the Thames estuary may not continue to provide a satisfactory solution, and other means of land and sea disposal are therefore being investigated. These include pumping through a pipeline to a point farther down the Thames and shipping from there to sea; pumping through a pipeline to an outfall in the North Sea; incineration (after de-watering) with or without household refuse; and spreading on land as a soil conditioner after de-watering.

Control of Trade Effluents

Toxic or deleterious industrial waste discharges could endanger men working in the sewers, cause structural damage to the sewers, or lead to the breakdown of plant and processes at the sewage treatment works, and strict control of such discharges is therefore essential.

The Council has sought to rationalise the control of trade wastes, and has introduced standard methods of charging for the reception, conveyance and treatment of trade effluents throughout its area. All accounts are now computer controlled, and it is hoped to use peripheral equipment to test samples and feed charging data into the computer.

In 1969–70 there were 5,700 controlled discharges, over 5,000 samples of trade effluents were taken, and the income due to the Council was £837,000.

Re-use of Effluent

The Council's sewage works produce 550 million gallons of effluent daily. Industry could probably use a considerable proportion for various processes instead of pure water, to the financial benefit of itself and the Council.

In fact, 5½ million gallons a day has been pumped from Beckton sewage treatment works to the North Thames Gas Board's works since 1967 and agreement in principle has been reached for the supply of effluent from Riverside works (Havering) to the Essex River Authority. Proposals for further re-use of effluent are being investigated.

Prevention of Pollution of Rivers

In the "London excluded area", an area of about 400 square miles which is excluded from the jurisdiction of river authorities, the Council is, in general, responsible for flood prevention, land drainage and control of river pollution. The Port of London Authority is responsible for the prevention of pollution of the Thames downstream of Teddington Weir.

None of the watercourses, for which the Council is responsible, north of the River Thames receives effluent from the Council's sewage treatment works; south of the Thames comparatively large quantities of treated sewage effluent are discharged to some streams. Other pollution derives from storm overflows from soil sewers, trade effluents, wrong connection of domestic soil drainage, and oil from garages, factories, hard standings and roads.

All so-called "main metropolitan watercourses" are patrolled regularly and frequently, and a sampling schedule, both of the rivers and of the discharges to them, is operated. The Council's riversmen regularly remove debris such as old bicycles, perambulators, and grass and hedge clippings. Sewage treatment works, trade premises from which discharges take place, and surface-water sewer outfalls are regularly inspected and sampled.

One of the most serious forms of pollution is from spilt oil, and in Greater London there are many places where oil is stored for one purpose or another. Ideally these should be protected by bund walls or some other device; often there is no protection, and spilled oil runs into the nearest drain. Where the discharge is to a surface-water system, miles of river may be affected.

Although, as already said, the prevention of pollution of the Thames is not the direct responsibility of the Council, since 1893 the London County Council and the Greater London Council have carried out daily samplings and analysis.

Besides the daily samples taken from the river at the Beckton and Crossness sewage treatment works, samples are taken at low water from Richmond to Southend and at high water from Richmond to the sludge-dumping area at the Barrow Deep in the outer estuary; and the Thames Conservancy makes available details of the average daily flow and an analysis of the water entering the river at Teddington Weir.

In order to assess the condition of the water at any given time and to compare its condition from season to season and from year to year, a full analysis is made, including in most cases the determination of chlorions (present as chloride); nitrogen (present as ammonia, nitrite and nitrate); chemical oxygen absorption; biochemical oxygen absorption; dissolved oxygen content; suspended solids and loss on ignition of these; and the acidity or alkalinity measured as pH. Other factors are the water temperature and the quantity of fresh water coming into the river at Teddington Weir, the latter of course being largely determined by rainfall.

Looking Ahead

Pollution is as great a problem of the age as the growing scarcity of clean water; yet water is increasingly used for the transport of polluting matter. New pollution problems arise while old ones are still being tackled. More and more rubbish and trade wastes are being flushed down drains. Since the Crossness extensions were opened in 1964 the condition of the River Thames has much improved and there will be further improvement when the works in progress at Beckton are finished.

159

Much still remains to be done, however, in improving and maintaining the performance of the Council's sewage treatment works, and in trying at the same time to reduce running costs still further.

THE COST OF THE SERVICE – 1969–70

Operations and maintenance

Sewerage and pumping	£1,726,000
Sewage treatment and disposal	2,673,000
Rivers (land drainage, flood prevention and pollution control)	304,000
Total operation and maintenance	4,703,000
Administration and rates	2,801,000
Debt charges	5,479,000
	12,983,000
Income	1,103,000
Net revenue expenditure	£11,880,000
Equivalent to one penny per day per head of the resident population.	
Capital expenditure during the year	£8,671,000
(Sewerage and sewage treatment and rivers only.)	

Pollution by Industrial Waste and Sewage: Legal and Administrative Aspects

Mr Ian Drummond (Clerk of the Trent River Authority)

A. Introduction

1. This is a short paper dealing with one of the great problems of our time -- pollution by industrial waste and by sewage. In this we are all responsible in greater or lesser degree. Certainly far greater education of all is required. But this is no substitute for sound systems of handling this problem, which is now assuming a new dimension.

2. In the next thirty years water use and effluents will double. The present system for controlling the condition of our rivers is clearly not working adequately and a far-reaching and fundamental reappraisal is required if we are to provide economically for our water resources in the future and to have rivers that can be fully used for amenity, recreation and the like.

3. This paper is largely confined to the river pollution problems of England and Wales. In greater or lesser degree, according to size of country, population

and size of rivers, they are fairly typical of problems in a modern urban and industrial society. Some reference is made later in this paper to systems used in other countries. It is emphasised, too, that this paper concentrates on the problem of sewage disposal, as 70 per cent of liquid industrial waste goes to the sewers and the pollution load of rivers attributable to directly discharged industrial wastes is generally minor compared with the load from sewage works.

B. Common Law

4. Control of river pollution at Common Law is based upon the property rights of the owner of land adjoining a stream ("riparian rights"). He has the right to have the water of the stream come to him in its natural state in flow, quantity and quality and to go from him without obstruction. Thus, if a person does anything to pollute or affect the natural quality of a stream, any riparian owner downstream has an action against him at Common Law for damages for loss suffered and may ask the Court for an injunction to restrain the offender from committing further acts of pollution.

5. This system of control may very well have been adequate in a non-industrial society where incidents of pollution would invariably be of a minor nature and localised in their effects. With the onset of modern society, however, involving vast concentrations of industry and population in urban areas and the development of water-borne sewerage systems, some form of statutory control was required.

6. That there is now a considerable system of statutory controls is not to imply that the position under Common Law is of no significance. Local sewerage authorities, in particular, are still extremely vulnerable to legal action by riparian owners. The most important case in recent years was Pride of Derby and Derbyshire Angling Association v. British Celanese and others (1953) in which riparian owners were awarded damages and a suspended injunction against, among others, Derby Corporation, in respect of the discharge from their sewage works.

7. It is noteworthy that Section 331 of the Public Health Act, 1936, specifically provides that the Act shall not authorise a local authority injuriously to affect water rights without the consent of the owner of those rights. It is not thought that such consent was always obtained and it would rarely be practicable to obtain it from all persons who might possibly be affected. There must be a vast number of potential claims by riparian owners which, either through ignorance or indifference, are not pursued. Clearly in a modern society riparian rights are impracticable and will have to be severely curtailed. What we have to do now is merely to license and limit pollution.

C. Statutory Control

8. Although there were provisions for the control of river pollution in the Public Health legislation of the nineteenth century, the first enactment to deal solely with this subject was the Rivers Pollution Prevention Act of 1876. The 1876 Act remained the basis of statutory control until 1951 but, in fact, it failed to contain the ever-increasing deterioration in the situation. Whilst the Act

defined and prohibited the principal types of pollution, the safeguards for industrialists and sewerage authorities and defences to prosecution were very far-reaching and enforcement was in the hands of local authorities which were too small and numerous to have the necessary expertise or regard to the regional aspects of the problem. A particular defect was that in so many cases the local authority areas contained only part of a river and often the local authority was itself a serious polluter. In some of the worst affected areas *ad hoc* bodies were established, endowed with special powers under Private Acts.

9. The present legislation was not initiated until after the Second World War. It began with the setting up in 1948 of thirty-two river boards which, together with the Thames and Lee Conservancies, covered the whole of England and Wales with the exception of the London excluded area. The new boards were established to administer the pollution control powers of local authorities together with other functions relating to land drainage and fisheries formerly exercised by river catchment boards.

10. In 1951, the Rivers (Prevention of Pollution) Act, having the express intention of maintaining or restoring the wholesomeness of our rivers, repealed the former legislation and created a new principal offence of "causing or knowingly permitting poisonous noxious or polluting matter to enter a stream" (Section 2). The Act also provided for all new or altered discharges of trade or sewage effluent to be subject to licensing by the newly constituted river boards (Section 7). In 1960 the licensing procedures were extended to new or altered discharges to tidal rivers and estuaries by the Clean Rivers (Estuaries and Tidal Waters) Act. The Rivers (Prevention of Pollution) Act, 1961, extended the scheme still further so as to require the consent of river boards to the continuation of pre-1951 discharges to non-tidal streams. Pre-1960 discharges to tidal waters are still not generally the subject of control by river authorities. Discharges to the open sea are completely outside the present system but the time is approaching when these too should be subject to control by the river authority.

11. The most recent legislation affecting the matter is the Water Resources Act, 1963. The main purpose of this Act was to establish twenty-seven river authorities in place of the river boards and to endow them and the Thames and Lee Conservancies with new and important powers relating to the management of water resources in their catchment areas, in addition to the functions formerly exercised by river boards. With the general aim of conserving and securing the proper use of water resources, the Act made certain amendments to the existing legislation on pollution control. Of some importance was a new provision in Section 72 to prohibit the discharge into underground strata by means of any well, borehole or pipe any trade or sewage effluent or any poisonous, noxious or polluting matter without the consent of the river authority.

12. It should be made clear that consents of river authorities to the discharge of effluents under the Acts may specify conditions as to the nature and composition, temperature, volume or rate of discharge of the effluents. Most river authorities have standard conditions for sewage effluents which are basically Royal Commission Standard but include additional restrictions and prohibitions of various substances which could be harmful to the river. It is pertinent to note

162

that Royal Commission Standard was recommended as long ago as 1916 and in 1964–65 60 per cent of the sewage works in the country did not comply. It is important to remember also that the standard was based on an assumption that the effluent would be diluted in a least eight times as much clean river water, and this point is often forgotten.

D. Inadequacies in the Existing Controls

13. Before dealing with the general question of whether the existing system of control is the right approach to a solution of the problem, it is proposed to consider briefly a few particular fields in which the system either is not complete or is difficult for river authorities to operate effectively.

14. It has already been stated that the powers of river authorities over pollution of tidal waters are strictly limited to control of discharges of trade or sewage effluent begun after 1960. Full control powers as they apply to non-tidal streams may be applied to tidal rivers and estuaries by order of the Minister of Housing and Local Government. Many estuaries in England and Wales carry a high pollution load from adjacent industrial areas (e.g. the Humber, Mersey, Tees, Thames and Tyne). River authorities have agreed that the system of control under the 1951 and 1961 Acts should be extended to the tidal waters covered by the 1960 Act. It is hoped that the Jeger Committee on Sewage Disposal, whose Report is expected this year, will make such a recommendation.

15. A principal difficulty in the administration of pollution is the right of owners and occupiers under the Public Health Act, 1936, to connect their premises to the public sewer. This right does not extend to the discharge of trade effluents, which will be dealt with later in the paper. A consequence of this right is that once planning permission in outline is granted, the local authority is bound to receive sewage from any houses erected. New housing is often a replacement character – for example, slum clearance. This does not create a problem except in increased volume of sewage because of the greater water using fittings in new houses, unless, of course, the replacement houses are erected in the catchment area of a different sewage works – with inadequate capacity. The problem arises, however, particularly where the occupants of the houses are immigrants, i.e. new to the area. The "flight to the fringes", i.e. from the centre of conurbations, has produced acute problems of this order in county areas.

16. It is, of course, the duty of the local authority under the Public Health Act, 1936, to provide such public sewers as may be necessary for effectually draining a district and to make provision, by means of sewage disposal works or otherwise, for effectively dealing with the contents of the sewers. Unfortunately it is not the habit of local authorities normally to maintain their sewage works in advance of the demand made upon them, in other words, they are very often overloaded and this is clearly shown by rivers which are polluted. A river is, of course, a form of safety valve for an overloaded works. But a power station or a gas works cannot be overloaded without serious consequences, nor can a waterworks provide more water than it in fact has.

17. The river authority remedy under the Pollution Acts is to prosecute for breach of consent – but this is a case of horses and stable doors. Accordingly, in

163

various ways river authorities have made arrangements with planning authorities for the latter, through their planning control, to curtail development until sewage works capacity can be improved. In many areas this has proved a not unsuccessful means of preventing overload or further overload of sewage works.

18. River authorities have recommended that in any new legislation provision should be made that no new development should be permitted without a certificate from the river authority that adequate sewage works capacity is available to deal with the increased load.

19. It is important here to consider specifically the case of industrial development. About 70 per cent of industrial waste goes to the sewers and consequently to the sewage works. An industrialist pays effluent charges to the local authority for their treatment costs. Discharges of trade effluent to sewers are regulated by the Public Health (Drainage of Trade Premises) Act, 1937, under which a consents system is operated by the local authority, subject to appeals to the Ministry of Housing and Local Government. The consents may impose conditions as to nature, composition and discharge of the trade effluent. Discharges of trade effluents lawfully made before the Act are exempt.

20. Trade effluents can cause many problems for a sewage treatment works (which is essentially a vast colony of microbes engaged in their normal lives, including gastronomic activities). Local authorities often have accepted trade effluents to their sewers for which their sewage works do not have capacity. There are naturally pressures for improving the local employment position and, of course, a new factory is enhanced rateable value.

21. River authorities sometimes endeavour to operate the same liaison arrangements with planning authorities in respect of industrial development as in the case of housing and non-industrial construction. For various reasons this has not proved very successful. On the other hand, a trade effluent wrongly accepted can damage a sewage works and consequentially a river. What is the use of granting industrial development certificates if the sewage works cannot handle effluent from the new industrial premises to be constructed? It is not always realised that trade effluents to sewers can represent 50 per cent of the total flow of effluent from sewage works.

22. There have been recommendations in the past that a river authority should be consulted about trade effluents which can have significant effect upon sewage works, but this is not often done. Accordingly, in this case as well, river authorities are recommending that in any new legislation no new development should be permitted without a certificate from the river authority that adequate sewage works capacity is available.

23. It may well be important, too, in the future for the river authority to know of new industrial processes which produce new compounds which could cause difficulty in rivers – perhaps for public water supply undertakers. This represents fair administrative difficulties. A substance which has not been created cannot be specified in a consent. On the other hand, it is difficult to frame a condition which covers the matter in general terms.

24. The discharge of surface water is a matter which has received little attention in the relevant statutory provisions. Apart from the obvious effects of surface water run-off upon land drainage and flood prevention interests, it also

presents pollution problems which hitherto have not been greatly appreciated. A heavy summer storm washing dirt and grime from the buildings, streets and drainage system of an urban area can impose a sudden grossly polluting load upon the river system. Salt spread on roads during the winter months can also have a very damaging effect on river water quality. Surface water discharges from a local authority sewerage system are somewhat inadequately controlled from the point of view of pollution – as well as land drainage. Additional control powers are clearly needed.

25. Finally there is the question of the protection of ground water and the limited provision for this in Section 72 of the Water Resources Act, 1963. The tipping of liquid or solid wastes on the surface of land can, by percolation, seriously affect underground water but unfortunately this type of activity is outside the scope of Section 72. Planning restrictions on these activities have not always proved to be effective and there is a need for direct control by river authorities if the quality of our ground water resources is to be preserved.

E. Need for Fundamental Re-appraisal of Administrative Machinery

26. Whilst the condition of many of our rivers is improving, even now some twenty years after the beginning of the existing system of statutory controls, there are many which are grossly or considerably polluted. The Trent, for example, while recovering, is still not of a standard suitable for potable water abstraction. If it is suitable, it would be equivalent in yield to a Wash Barrage. The Trent River Authority are co-operating with the Government in a major research programme coupled with a river economic model. The results of this programme are expected next year and it is hoped that they will show the most economic way of meeting the demands on the river system from the resources available.

27. Where water resources are concerned the management of water quality is just as important as the management of quantity, in fact they are one. While the management of the latter is fairly reliable through abstraction consents, the same can hardly be said of the former. With the size of towns and their industries today, effluents from sewage works are often large in relation to the rivers into which they discharge. They could well be described as "Swords of Damocles". In the next thirty years the total of effluent is likely to double with the appropriate requirements of increase in sewage works capacity and of higher effluent quality standards, since the natural flow of rivers is hardly likely to increase.

28. In England and Wales potable water at present comes from three sources, as follows:

one-third – uplands, lakes and reservoirs;
one-third – ground sources;
one-third – direct from rivers.

Water demand will double in the next thirty years. Where will it come from – leaving aside such matters as desalination or barrages? A doubling of reservoirs is an interesting political proposition. A doubling of ground water seems hardly credible. Can you double the take from rivers when they are often polluted and when there will be twice the effluent in them in thirty years' time? The new

165

reservoirs to be built will be the river regulating type where the river is used as an aqueduct. Further water is proposed to be moved from one river catchment to another using river channels. All this requires good water quality. Obviously a high integration of management of rivers and all forms of water installations including sewage works is needed and an investigation of these matters has already been put in train by the Government.

29. Most industrial countries of recent years have indicated similar problems in this field, and are naturally concerned to establish the best possible administrative solutions for the effective integration of the water conservation planning and development on the one hand, and the control of pollution on the other. The recent Seminar on River Basin Management of the Economic Commission for Europe clearly bore this out. Integrated river management was a dominant theme of the Conference.

30. We now await the Report of the Ministry of Housing and Local Government Working Party on Sewage Disposal (the Jeger Committee). The Ministry, with the aid of the river authorities, are also carrying out the first complete survey of river conditions in England and Wales since 1958 and the results of this will become available early next year. In 1969 the Government re-established the Central Advisory Water Committee to advise it on the reorganisation of the water and sewage industry. As the White Paper on *The Protection of the Environment* (May 1970) indicates, the industry will, in any case, have to be reorganised at the same time as local government. Any new distribution of functions in the control of fresh water pollution will be part of this major restructuring of the industry. Finally, the Government have established the standing Royal Commission on Environmental Pollution. They will be visiting the Trent River Authority in October next. As the White Paper indicated, structural re-organisation of a water industry, improved techniques, higher standards, more money and more comprehensive legislation will make possible a significant improvement of the condition of our streams, rivers, lakes and reservoirs during the 1970s.

31. In this general connection the following statement of the Water Resources Board in their recently published Sixth Annual Report is of interest:

"But we believe that the basin planning authorities must have a mandate to optimise the pattern of storage works and similarly to exert a positive influence on the grouping and location of sewage outfalls and the capacity and nature of the treatment plant installed. To a considerable degree the needs of flow regulation, pollution control or mitigation and, for that matter, river channel maintenance for flood disposal, are interdependent; and this should be reflected in the administrative and financial arrangements for dealing with them."

32. Naturally many solutions to the problem of reorganisation of the water and sewage industry given to the Central Advisory Water Committee by the Government have been published by various bodies and persons. Three are set out below:

(a) Closer co-ordination of existing public water supply undertakers and sewage works authorities by river authorities operating through a comprehensive plan;

(b) The same as (a) above with the variant that Joint Sewage Boards should be established to manage sewage;

(c) The management of sewage works by river authorities. Under this solution public water supply undertakers could be integrated into the design or otherwise. Finance for sewage treatment under this solution could be raised through a charging scheme based on the cost of treatment according to volume and quality of sewage.

33. A major complaint against the present system is that while the local authority benefits from the conduction of sewage away from urban premises, it does not benefit from sewage treatment. This is a largely non-urban benefit. Further, the system is regulatory and cannot easily encourage a system of complete co-operation and disclosure of problems. The local authority is "at risk" as proceedings can be taken against it. It is not possible to consider economic choices of action based upon freedom to vary quality, flows and the like. These defects apply in greater or lesser degree to (a) and (b) above. In the case of solution (c) there is effective integration of the major elements which have effect upon river systems. The sewage works manager becomes a pollution officer for his own works and there is complete knowledge of treatment problems down the system. The solution is, however, open to the objection that the river authority is both poacher and gamekeeper. It would, however, control by far the greater quantity of effluents going into its system as only about 30 per cent of industrial effluents do not go to the sewer and this figure is becoming less. In terms of pollution load, certainly in the Trent area, sewage works effluent represents probably 85 to 90 per cent of the pollution load of the system. The national figure is probably not much different. The defect here could be remedied by giving a power to the central body such as the Water Resources Board, to take action against any river authority that was in default. Further, it would indeed be a brave river authority that complained about an industrial effluent when its own hands were not overclean ("he who comes into equity . . .").

34. To conclude, it is a far cry mentally from the start of this paper with its account of early pollution controls. Present trends in improvement of administration show a clear understanding of what, of course, has always been there, namely the natural system of a river basin now increasingly being dictated by man. The only hope for its future health and use by man is comprehensive management. In this, industrial and sewage wastes are part only of the affair though albeit a most important part – not only as sources of considerable danger to the system but also as sources of considerable wealth. The treatment on commercial lines of sewage may not be possible but its throw-aways built into a soundly managed river system can be of considerable economic value to the community. It would be useful, particularly having regard to the objectives of the Institute, to consider what other countries of the Western European type have done. Pollution control as we know it in this country is operated generally and is of a regulatory character. There have, however, been some interesting examples of integration of functions of river basins of the type that is now being considered at home.

The genesis for this approach is related quite often to the population density of the particular country, the size of its rivers and the proportion of the population that are on main drainage. England and Wales have one of the highest population

167

densities in the world and only 3 million of its 50 million are not on main drainage.

The Ruhr river authorities – some of them founded in the early part of this century – are famous for their multi-purpose nature and their success. They handle land drainage, pollution, reservoirs, sewage works and river aeration lagoons. They have an area of 4,300 square miles containing 10 million people and they have developed comparatively sophisticated methods of distributing costs of their water quality operations by levying charges on effluents received by their system.

France has recently established river basin authorities and reinforced existing pollution control. Their river basin authorities have power to construct and operate sewage works. Charges will be made on effluents to finance these works. It is interesting to note that her population density is much less than this country and less are on main drainage.

The Ontario Water Resources Commission is an interesting new development formed in 1958. It is a multi-purpose river basin authority which constructs and operates sewage works.

Finally, the United States (even with their enormous area and low population density) has recently had to introduce more effective pollution control of a regulatory character and establish river commissions (which are at present largely planning authorities). There are, however, examples of comprehensive river basin management authorities including power to operate waste treatment plants with the Tennessee Valley Authority and Delaware River Basin Commission.

Discussion

SIR FREDERICK WARNER, CHAIRMAN
We shall begin our discussion with any questions from members of the Conference to the two individual speakers, before we take up general points.

PROFESSOR J. F. GARNER
Could I follow up a point from Mr O'Sullivan? He was talking about the right to discharge to the sea. I think the Common Law, anyway, does not recognise such a right. For example, there was a nice case in 1906 when the Mayor of Chichester gave an oyster feast and large numbers of people felt very ill afterwards, and it was found that it was due to the Warblington Council's sewage works polluting the oyster fishery. The owner of the oyster fishery was able to sue for damages, although the pollution was in the sea outside the river itself. Then, of course, there is a famous case, a little while ago, when a tanker broke its back in the Ribble and oil escaped and came on to Southport Corporation's bathing beach, and the Southport Corporation was able to recover damages from Esso, the charterers of the tanker. So there are possibilities of using the Common Law when we are talking about effluent to the sea. I also wanted to say a word about "detrimental". It seems to me that it is a much better word in this context than "poisonous" in the river legislation. "Detrimental" – I quite agree with Mr O'Sullivan – I think can be construed in a wide way as he suggested. And now I finally wanted to ask a question of Mr Pearson about this sludge business, as to whether, really, it is not a waste to send it out to sea?

Mr R. E. Woodward

I suppose I come under the heading of an administrator or a manager. I happen to be a lawyer; I am not a scientist. Could I just say this? We have been concerning ourselves in the past two days, particularly today, with this question of the prevention of pollution and I am persuaded, more and more, that less and less is this a technological problem. It is a problem of economics, finance, politics – herein lies the problem. Mr Drummond, my neighbour and friend, has shown you a map of the Trent River System, illustrating the extent of pollution. If I were to show you a similar map of the Mersey and Weaver system, his would be paradise. But, you know, between the two of us, we serve a population – and I am making a guess – of about 10 million. There are $5\frac{1}{4}$ million people in my area and I apprehend that this is rather more than that in Mr Drummond's – making about a fifth of the population of the country. I want to reiterate and emphasise, as best I can, what Mr Drummond has said to you this afternoon. What is urgently needed is a re-orientation of the management of water conservation in this country. I have listened, silently, for two days and a good deal of the dialogue has been directed at industry, but I was so pleased when Mr Drummond emphasised the question of sewage disposal, because there are only two types of effluent. Effluent is either industrial effluent or sewage effluent. If I could just leave you with this thought – and this is not to let the industrialists off the hook here in the slightest – it is trite but true that there are no votes in sewage, and the sooner we get the financing of sewage disposal out of the general rate fund the better, because there is certainly no motivation to improve our rivers so long as Local Government controls sewage disposal. I think it is important, too, to make the point that it will not be any better, in my submission, when you get the reform of Local Government, and for this reason: that there are in this country today about 1,500 sewage disposal authorities, and, come Maude or the modification of Maude (which is inevitable), the number of sewage disposal authorities will be reduced down, perhaps, to two figures. But where will the motivation be in those authorities to put sewage disposal at the top of the list, against housing and planning and so on? I suggest the situation will be the same; there are, as I say, many facets of this problem. I do not want to take up any more of your time, I just want to emphasise this point; that I think that, in relation to sewage disposal, somehow or other we have got to be motivated away from Local Government. Another thought: have you ever asked the average local councillor whether he has ever seen his sewage works? Or course he has not! He could not care less! Sewage effluent goes out of the pipe, into the river, and he says "Ta Ta" to it – and it is at that point, under the present legislation, that the River Authority takes over – they take what somebody else gives them. The motivation is all wrong; whatever else we do in the reorganisation of water conservation, we must put sewage disposal into the hands of the people who are concerned with the effluent that goes out of the pipe.

Sir Frederick Warner, Chairman

I wonder if Lord Simon would care to tell us how his River Authority deals with 10 million people (which is the last speaker's Authority and the Trent River Authority combined), and has managed to get the Greater London Council to spend the amount of money which Mr Pearson says they have spent. Where does the motivation come from?

LORD SIMON

I recognise very well that I speak as a complete layman – I am an administrator, I am certainly not, in this context, a scientist or a lawyer – and your Chairman could tell you a great deal more about this than I can, and so, I am sure, can some of my friends from the Greater London Council who are here. I was very much interested by what Mr Drummond or somebody else said about the necessity for co-operation in this matter. It is quite interesting that the Port of London Authority became the Pollution Authority for the tidal Thames in 1909, when it was set up, in the time of the London County Council, but, owing to the great power of the London County Council – very properly, at that time – they were actually exempted from the legal control, so that all the work that was done, until a few years ago, in improving the outfalls of the London County Council and, later, the Greater London Council, was done by persuasion, and not by any legal powers at all. In this case I am bound to say that, it seems to me, that provided the Pollution Authority and the Local Authority are of one mind, a lot of progress can be made without any sanction. As a matter of fact, the sanction was introduced about three years ago, very much against our wish. The Government of the day said it was quite improper that the Greater London Council should not be subject to the same rules as everybody else, so sanction was introduced. We met that position by setting up a Pollution Committee on which representatives of the Council sit, and we still discuss these matters round and not across the table.

Somebody earlier on talked about losing the advantage, as he felt, of the policeman or, perhaps, the gamekeeper–poacher attitude. I, personally, am completely opposed to that. I think as soon as you get the gamekeeper–poacher attitude, you get people defending their own position and you do not get co-operation, and I am very happy indeed that we have not got that feeling. What Mr Drummond has proposed would, of course, get over that in another way; that is, by having a single authority responsible for the sewage works and for the effluent that came from it, and I am not arguing against that, although I still think that there are problems in it, because in any organisation which has a number of functions to perform, there are people, individuals, men and women, who are in charge of particular departments, and we know, even in one big organisation, you still have fights between the departments as to who is going to get the money to do this or that. But, basically, I agree with Mr Drummond, and, let me say at once, my first feeling on reading his paper was that I did not, but I think I have come to agree with him, it is probably the best solution. But I can say, as I was invited to by the Chairman, that we have been able to effect a great improvement in the condition of the tidal Thames by co-operation, by working out together the steps that ought to be taken, the successive standards which we should seek, by setting a standard that, I am bound to confess, we have not yet reached but we are advancing towards it, and, very largely thanks to our Chairman and his colleagues, by having behind us a really thorough scientific study of what the problem was and how it was best answered.

I can say one other thing before I sit down, and that is I listened with great interest, of course, to Mr Pearson's exposition and also read the paper that he had given us before, or has just circulated, and I do want to say, because I think it ought to be said, that the Greater London Council and their predecessors, the London County Council, have all along been very conscious of their responsibility in this matter. There may be no votes in sewage, but they have gone ahead with dealing with the problem, votes or no votes. I think one should, perhaps add this: that a very large local authority such as this, where a considerable number of the voters do, in fact, live or work on the banks of the river, has, of course, got a little more incentive to improve the position than some who are a long way away, and I can think, and I dare say Mr Pearson can think too, of one works which gave us a great deal of trouble, where the inhabitants and councillors were in one location, and they were discharging into the river and not one of their voters lived on the bank of the river, so they were quite unconcerned with what happened there, and this re-echoes what somebody has already said.

Mr F. Macdonald

Could I speak on this complex of control, cost and co-operation? Perhaps my presence here today will demonstrate this. I have been nominated to come here by the Confederation of British Industry, my expenses are paid by my company, and I would like to talk on behalf of the River Authority. I think this is an example, but what I would like to bring to your attention is that co-operation can stop at any time. I think, looking through the Association of River Authorities, the total cost of the River Authorities was about £22 million in 1969, and, if you add the £3½ million from the Greater London Authority, it is £25½ million, but this covers six categories: water resources, etc. If you look at the end of the line, at pollution prevention, that got about £2 million – I may be corrected, but I think this is around the figure. It was said this morning that the River Authority is the policeman and then the legal people said if you give us the facts we could then take action. And then, I think, a nanagram was mentioned – a nanagram is a million-millionth of something which it was necessary to estimate – and then arising from this necessity there would be high expenses for this type of test equipment for PCBs. I think on PCBs there are perhaps only three instruments in this country that can estimate them correctly. I am chairman of a Pollution Prevention Committee and all the local councillors have seats on this authority, and when I put my budget forward at the end of the year, we get pruned and cut and pruned and cut.

I am on the Severn Estuary, along with Mr Whitfield – though higher up – but the Severn Estuary alone, with all the loads that are going into it, requires a tremendous amount of data and fact finding, and this is an expensive business. I think our outlay on the Usk River Authority for pollution prevention last year was £27,000. Some of the required equipment alone will cost this sum, so somebody has got to take over the responsibility. Grant aid will buy you the equipment, will buy you instruments, but this is not enough. Somebody has got to take money control from the local authority because pollution is a problem which is national and international.

171

Mr G. W. Hull

May I please underline what the last speaker and the Mersey and Weaver River Board representative has said? I attempted to raise the question of municipal waste disposal at OECD about seven years ago without success. This is a common national problem, and so are the costs of other environment research which could be reduced by international co-operation. Clearly, Mr O'Sullivan and several others can see much research work that should be done. But the only way in which international co-operation on this subject is likely to be achieved seems to be by strong pressure on national governments. There should be a sufficient number of representatives of influential bodies here to do this, to list the matters we have been discussing here, and put them forward to government through their own organisations, with a view to research, economic co-operation, and countrywide recommendation.

Mr J. C. Hanbury

Can I deal with one small point? I have been waiting for someone to comment on the fertiliser value of sewage sludges, because there is much misconception about this – the potential value of such sludges tends to be greatly exaggerated. I do not carry a lot of detailed figures in my head, but there are two or three serious snags. First of all, the high cost of processing sludge to a condition in which it can be used; secondly, sludge is – we must be frank about this – a poor fertiliser, it is out of balance, and there is practically no potash in it, and it has limited value unless it is scientifically formulated to make good the deficiencies; thirdly, the total quantity available is surprisingly small. If every ton of potentially usable sludge were, in fact, used in agriculture – do not take me too literally on this figure – it would still only amount to about 7 per cent of the total tonnage of nitrogen required in this country, something of that order.

Dr D. W. Bowett

Yes, I think people looking at sludge ought to be aware that sludge, as it comes out of the sewage works, is 99 per cent water, and when it is dried it is 95 per cent water.

Dr W. R. P. Bourne

I think one of the points which emerged from the Royal Society meeting on the Pollution of the Sea, earlier this year, is that the reason why sewage cannot be put back on the land as it used to be is the amount of indestructible residues in it. I have not brought up the bird figures yet but when these things get into the sea – we are at the extreme end of the chain. After the *Torrey Canyon* we started counting the auks, because there was obviously something happening to them. Between 1967 and 1969 guillemots in the Irish Sea area went down by a fifth. We thought this was due to oil pollution until last autumn, when we found 15,000 dead, of which only 15 per cent were oiled. Well, these were only the ones we found; this year we have been counting the breeding colonies again, and in the half that have been counted so far, compared with last season, there has been a decline of another 50 per cent, and razor-bills, which were formerly holding their numbers, have gone down by 30 per cent, which implies that the total mortality must have been several times as large; this is a very big hole in the population. It is not due to oil pollution. If there is a natural explanation, one wants to know what it is, because nobody has yet provided a natural explanation. There is a big question hanging over the PCBs which has not yet been resolved, and it is not the question of a single river which is polluted, but quite a major part of the sea; the whole area from the approaches of the Firth of Clyde to the south shore of Devon, and I think this is the beginning of the long-term repercussions from putting things into the sea.

SIR FREDERICK WARNER, CHAIRMAN
Would anyone like to comment on Mr Woodward's point (it really is related to the last point) that he could see the only obstacles to dealing with the problems as lying in the economic and political field rather than in the technical. Is everyone convinced that all the technical solutions are available?

PROFESSOR E. EISNER
I do not think it is accurate to say that it is not a technological problem, because what the economic problem is is determined by what the technological solutions are. If the given technological solution makes something much cheaper, it may alter the whole economic picture. Furthermore, another aspect where technology is very seriously involved, is that the economist and the politician need to know the relations between cause and effect before they can make a decision, and determination of the cause and effect relationships is very much a scientific and technological matter.

MR J. F. WHITFIELD
I would like to refer to remarks of my contemporary on the Usk River Authority. I think, basically, and it is only my opinion, that the main problem of river authority work and management is essentially political. The technological and legal problems that present themselves, ultimately, in general get solved. I feel that the whole question of river authority overall management needs a complete review. As a great many of us here feel and as I have already expressed, one of the greatest weaknesses, politically, in the present structure of river authorities, is the built-in representation of local authorities, which, as you know, provide money. They have 51 per cent voting power in the current river authority structure and, therefore, they have the say in precepts. We river authority people have got to go with a begging bowl and, as has already been very ably expressed, we have got to cut our budgets down and down. Local authority representatives are there, some not for the benefit of the river or of the related authority but for the benefit of their own local council in cutting down rates, and they do so quite effectively very often. The quality of membership on local authorities – I may be treading on some corns here – could be improved, generally speaking. I would say that the special representatives – from agriculture, industry, fisheries, land drainage, navigation, etc. – generally speaking are pretty good. The local authority representatives change over so many times with the politics of their own local committees that they, in many cases, act as a deterrent to the aims of the river authority.

PROFESSOR P. F. WAREING
It is very noticeable that all the representatives of river authorities here are in

173

favour of integrated control of water quality and water conservation, and yet the one local authority speaker this morning seems strongly opposed to this proposal. It seems that many local authorities are opposed to the control of sewage treatment passing out of their hands. I cannot understand why they should have this attitude. If there are "no votes in sewage" and if it is largely a charge on the rates, why are they so strongly opposed to the transfer of these responsibilities to a new authority responsible for both water quality and river management?

Mr J. C. Hanbury

With your permission, Mr Chairman, I simply must give a plug to the Jeger Committee Report. I am very sorry that it has not seen the light of day in time to be available for this discussion today, but I hope it will be available within a matter of days. I would like to say what this Committee has done. It was set up specifically to examine sewage in its widest and most comprehensive aspects and what it has done is to examine everything to do with water pollution, that is, rivers, estuaries and the sea-shore. I think you will find when you are able to read that Report that it really has identified the problems and the shortcomings in the present situation, and what has to be done to put them right. In effect, it says, "We have done our job, we have identified the problems and suggested the solutions that we think are the right ones. Now, it is over to you, the politicians, that is to say, the Government, to decide what action ought to be taken." The theme song is, "There have been words, words going on for a hundred years, and now it is time that somebody did something about it". On the organisational side, which we have not seriously attempted to deal with, apart from a few odd comments, we know that CAWC has made that its principal objective and I hope, by next year, that CAWC will be issuing a report with advice as to the future management of the whole of the water cycle, defining the role of central government, local government, sewage authorities, river authorities, the lot, comprehensively, and I think if we can possess our souls in patience until we know what they recommend, then we will really have something to get our teeth into.

Mr E. H. Hubbard

I would like to ask a question arising from what Mr Pearson said about the financial and technical problems of coping with the London water supply from now until the year 2000, and two figures which he put in the beginning of his report. I made a quick calculation which shows that the water consumption per inhabitant in London is 70 gallons a day. Now, I do not drink 70 gallons of water a day, gentlemen, it is impossible. Much of this goes in washing and the largest domestic part is what used to be called the "television flush". In fact by far the largest part is for industrial use. Do we need drinking water for that? Has anybody looked at the calculation, financially, of supplying two different types of water, one the larger part which is non-drinkable?

Mr R. Pearson

The total quantity of water consumption just quoted includes the water that comes down from industries as well as consumption by people. The average quantity used domestically throughout London varies. In some places it is down to 40, at other places up to an average of 60 gallons per head. The additional amount used is industrial.

Mr E. H. Hubbard
Would it not be true to say that the quality of water required for us to drink is probably higher than the quality of water required for industrial cooling and for me to flush my own toilet?

Mr R. Pearson
Indeed it is. But this is a well-known problem. The difficulty is that you have got to re-pipe the whole of London before you can provide a dual water system.

Dr D. W. Bowett
Mr Hubbard wants us to have something like France, where you just take the pipe system and use it for all purposes except drinking, and you have bottles for your water.

Dr W. G. Marley
Could I just follow this point one stage further? To what extent could water demands be economised by encouraging, very strongly, re-circulation of water in large industrial plants? That means that the plant would have to have its own purification plant for its own contaminants and then use the water over and over again, instead of just putting it into the sewers. In some instances, in the radio-activity field, we have practised this in laboratories, circulating the water, using it for cooling water and purposes like that where pure water is not necessary, and this obviates a lot of problems in treating the water so that it is fit to discharge.

Mr F. Macdonald
There is a very strong incentive to do this, I should have said. Well, I will leave it to you; it is 3s 6d per 1,000 gallons, roughly, to buy your water and, if you have to dispose of it as trade effluent and you are charged by the Monckton Formula, then you find that the cost that you are being charged by the local authority for disposal will vary between 5s (the least) per 1,000 gallons up to, perhaps, 30s per 1,000 gallons. The economic incentive to recirculate water inside industry is already very great.

Could I give some figures? Actually the usage figures I looked up quite recently: CEGB took 5,000,000 million gallons of water in the year; industry, by direct extraction licence, took 1,667,000 million; and domestic, the water undertakers, also took a similar figure, about 1,600,000 million. What you have got to realise is that, of the 1,600,000 million the water undertakers took, a vast quantity goes to industry. So, I think, your domestic situation takes a diminishing value in this situation.

Mr J. R. Stribling
There is a great deal of thought about this subject, of course, and, as our friends over here said, why should it not be that waters which are used for cooling purposes and other industrial purposes which can be recirculated, be recirculated? Well, we have put this to industrialists over the years and the answer has always been: "If we pay £5,000 a year for our water bill, our shareholders know all

about it. If we include in our next year's accounts a bill for £50,000 for a re-circulation system, we shall get shot." And so, we have got very little answer on that point so far. But what the current thoughts are now, is that if any indus-trialist uses potable water for a non-potable use, then that should attract a very high charge, and this is now being applied. As a result our friends in industry are now saying, "If we cannot recirculate, why should we pay a very high factor charge for a quality of water we could do without? Let us try and obtain a second or third quality water for the purpose that we require and, therefore, only pay a second or third quality charge." This, I think, will be the answer.

MR D. H. THOMAS (Southampton – Rechem Ltd.)
I would like to raise two points against some of the arguments made here this afternoon, on the subject of recirculation of water.

Industry as a whole requires in most cases, and in particular the electronics section, a high purity water; in recirculating large or small volumes, it can only be done two or three times at the most, as any more than this creates a build-up of elemental metals and salts, just the same as we are getting with pesticides, a continuing build-up in polluted areas. So industry has to either get rid of this polluted water or treat it to remove the contaminants.

The second point on this same subject, is that the equipment for recirculating this water for industry, under modern accepted methods of technology, is ion exchange, and to do this efficiently, this equipment has to be regenerated with an acid or caustic. Unless treated properly, this regeneration alone will create further pollution problems.

So careful thought and study should be given, before talking of recycling water as if it was an easy operation, and the answer to the problem.

MR F. MACDONALD
Could I come back again on the question of recirculation: most of the industrial water that is used is for cooling purposes. We have the figure of the CEGB as a typical example that it is possible to recirculate. Processed water *can* go down the drain, but there is a certain amount of make-up water which can compensate for that. If you pay 3s 6d per 1,000 gallons, or 5s 6d as it will eventually go to, you can recirculate this for cooling water – perhaps 95 per cent of your usage – for about 5d or 6d per 1,000 gallons; I think you will find a return on your investment. We actually use 1 million gallons a day make-up in the plant that I am in, and we circulate nearly 2 million gallons an hour through a recovery and recirculation system. So it is possible.

MR B. M. ARCHIBALD
I would like to confirm the remarks of the last speaker. It is very dangerous really to generalise too much on this subject. In the chemical industry, where a great deal of energy is used, and a great deal of water is required, not necessarily for cooling, it is certainly recirculated through, generally, the natural draught towers that you see in any large factory or refinery. They recirculate the water roughly twenty-four times, in other words, there is a 4 per cent make-up to offset what evaporates from the top of the tower – that is a pretty economical use. It may seem that the purity of that water is not important, but, in many chemical

processes, it *is* important. It is important, first of all, because you do not want to clog up your equipment with dirt, pollutants, salts in the water, which might be deposited in equipment, and necessitate the plant's being shut down more frequently for cleaning, perhaps difficult cleaning. It is also important because, in these plants which operate for a long time, leaks develop, sometimes one way, sometimes the other. If cooling water passes through a leak into process and is not of town potable quality, that can be very deleterious. Another major use of water, of course, in industry, all industry, is for raising steam. This has to be of a very high quality and, therefore, is treated by industry, but, if you start at potable standard, then you have an advantage. It is more economical, I would imagine, from a national point of view, to treat the water to potable standard on a large scale, supply it both to domestic use and to industry, and then let the industry work it up from there to its own particular grade or standard, than to instal a second system, distributing a lower grade water and have work-up at a great many different centres. As a rule, that is not a very sound thing to do.

Dr J. W. Hopton
Experience tells us that some biological effects of pollution are insidious and do not become manifest until twenty or thirty years after the event. Consequently we must consider long-term costs – it is conceivable that if we do not spend £1,000 on investigation and prevention now we may be faced with a health bill for £1 million in thirty years' time.

Dr J. A. G. Taylor (Unilever Research)
This business of water recirculation: I think it should be stressed that this is not a panacea by any means, because we still have the fact that there is the same amount of pollution eventually to be discharged and a lesser amount of water. This is not totally true, but, as a generalisation, it is fairly true. In a number of instances, we have tried water recirculation, and we do use it, but it is not always so easy. In one case, by doing this, we replaced a water pollution problem by an air pollution problem, and the local hospital certainly did not like it.

Mr P. Armstrong
I want to come back to this question of pollution by going into the production of polluted water. That is one thing. The other thing is, I think, that much of mankind is now putting its hope in what is called "the Green Revolution", that is to say the use of all these modern better seeds, and also pesticides in under-developed countries. I would like to know whether the shadow of that progress has been taken into consideration. That is to say, when these new plans are made for growing bigger crops in India, what happens if, in fact, their rivers, which they use very much more directly than we do – for drinking, and other things – are filled with these new substances? Secondly, a number of people have said that this is really a political question. How are you going to establish your priorities for putting money into this, and how are you going, in fact, to control it, how are you going to accept the fact that a lot of the pollution is going across frontiers, that coastal zones are the main pollutants, and it is coming out not merely from the rivers but, as Mr Hanbury was saying, from the seaside towns? For instance, you have all heard about Bognor – according to a witty MP you

cannot swim there, you can just go through the motions – and it is a fact. Can the international community really "get enough steam up" to take the essential decisions necessary to control these problems? The UN agencies – and there are a number of them – are, of course, seized with all this, but to come back again to the question which I asked earlier yesterday; if the political machine with its nationalism is so strong, whether in fact these multi-national corporations which are producing so many of these potentially progressive, but pollutant, substances, could be encouraged to set up a sort of self-policing system. Otherwise, it seems to me, there will be tremendous catastrophes, not from anybody's fault, but by mistake, although in a sense not just accidental because of the whole machine, the whole progress of mankind, is just careering along and nobody is in control or even thinking about the results too much.

Dr G. P. Hekstra

Mr Hubbard raised the problem of separating two systems, one for drinking purposes and one for other purposes. Would it not be possible to tackle the problem from the other end and make two systems for sewage; one for toilet waste and another for domestic wastes? The bulk of sewage is not from the toilets but from all other purposes within domestic use, so would it not be possible to separate the toilet system from all other sewage, especially in those new housing blocks in cities which are now being built? We have to tackle the problem for the future. We cannot, of course, now rebuild old cities, but we can introduce this system with the new blocks we are now building. For all new blocks of at least four storeys, or higher, it would be very easy to separate the toilet waste from all the other domestic wastes. This has already been done, as far as I am aware, in some places in Sweden. The sludge has been shown to be very good for manuring purposes, and it will make the sewerage system much cheaper, because the treatment of the other sewage will be much easier to carry out.

Mr R. F. Pearson

I think Dr Hekstra is referring to the separate system of sewerage. This is what applies in most new towns now; that is, parallel sewers for parallel sewerage, and separate sewers for surface water run-off.

Dr G. P. Hekstra

Water from baths should be separated from water from water closets.

Mr R. F. Pearson

I should say that this is a scientist's problem. I should have thought that water from baths is pretty well as difficult to treat as water from water closets. It has not got suspended solids, perhaps, but it is very difficult to treat, and I certainly would not want to put it into rivers. You would then have to have a twin purification system. Or a triple system; separate for rain water, separate for baths and a third for water closets.

Mr J. C. Hanbury

I happen to have seen the system that Dr Hekstra has spoken of actually working. In fact, the working party had a high old time one day in the yard of the Ministry

178

of Housing, when we had a mobile van with a mobile lavatory – or two mobile lavatories – and we played with this thing all morning. This is a commercial project and I must avoid doing a commercial for it, but I have seen the system working and I was greatly impressed with it. Speaking for myself, the two great features about this are, first of all, it involves a great economy of water because it flushes a lavatory pan efficiently with 1 litre of water instead of 2 to $2\frac{1}{2}$ gallons. So just think what that means in the saving of water; and, secondly, it involves separating what they call black water and grey water. "Black water" is what comes out of the lavatory pan, "grey water" is everything else. This does mean two separate purification systems, but it is very well worth while doing, because the lavatory water is, by bulk, a very small proportion of the whole; the rest of it does not need such drastic processing. I think I am right in saying – I am subject to correction – that one of the beauties of this system (pollution and its problems are an art and not a science as is medicine) is that it has been shown in practice to work on anything from one single house to a town of 40,000 inhabitants and maybe more. So, this is not just a van in the yard of the Ministry of Housing, this is a system which is working and, I understand, working well. It would be more difficult to extend this operation, I think, in this country than in Sweden – conditions are so different.

Dr G. P. Hekstra
But in future for the new blocks you could start the system in an entirely new area.

Mr J. C. Hanbury
Starting from the ground up, yes, that is so. I have often thought it should be done in new housing estates, council estates, but I am afraid I am getting into commerce, which is rather improper, so perhaps I had better close.

Lord Simon
I was wondering whether I could ask if perhaps Mr Pearson can help us. I thought – unless this is a completely new system of treating sewage – that in the sewage works it was very essential to get a good dilution of what we may call "water closet sewage" in order that the works would work. Is there some new system of treatment, which can deal with this particular section by itself?

Mr R. F. Pearson
I think that would be a scientist's job. If you take the domestic sewage out – which is what it comes to, for this is where all the organic matter is – I am not quite sure what the result would be. You then get more or less a trade effluent left to treat without any ordinary domestic organic matter with it. So this seems to be a different problem. I think it is a job for the scientists, in the first instance to tackle.

Sir Frederick Warner, Chairman
Is there an expert in the house?

Mr A. J. O'Sullivan
Treating a trade waste by itself can present many difficulties because constituents such as heavy metals, plastic waste, and other non-biodegradable substances

require special treatment processes. They will pass unchanged through the ordinary biological treatment processes or else are incapable by themselves of supporting the bacterial flora upon which biological treatment depends; this is because the bacteria require for their growth a certain range of substances such as nutrients, organic carbon and a number of trace elements, all of which are found in domestic sewage. Thus trade effluent which is biodegradable can be much more easily treated when mixed with domestic sewage. Now the suggestion that I made briefly in my own paper was that the conservative or non-biodegradable substances which are not affected by biological treatment and which pass straight through the rivers and eventually to coastal waters should be removed by pretreatment at the works where they are produced. Materials such as chemical waste, effluents from mothproofing, plastics manufacture, and heavy organic chemicals would fall into this category. If these were treated first so as to render the treated effluent "grey" rather than "black", this "grey' effluent would give rise to far less problems either at the sewage works or in the receiving water.

SIR FREDERICK WARNER, CHAIRMAN

I think you have had your answer already from Mr Thomas. You introduce another problem, because the only way you can take up chromium and things like that is by ion exchange. Then, in regeneration, you have got the problem of what you do with them. We still have not got to the point where anybody can tell us how the system of Mr Hanbury's works, nor on a very small scale how to turn out something as efficiently as does the general sewage works.

MR R. F. PEARSON

I am afraid I cannot really agree with Mr Hanbury nor with our friend from Holland because one of the problems is that if you think about separating the "black" from the "grey" you probably overlook the fact that the architect, when he builds this ten or fifteen storey block of flats, has, correctly and rightly, installed garbage grinders in every kitchen, and he has done this to prevent blockages of the drainage in the blocks of flats, and, therefore, you find that your so-called "grey" stream is much "blacker" than the "black". Taken from the organic point of view the fear of degrading from the so-called "grey" is so much more than from the "black" that I think you are wasting your time. Secondly, I think, too, it is practically impossible to re-drain all the existing towns in this country.

SIR FREDERICK WARNER, CHAIRMAN

Does it or does it not work in this supposed town of 40,000 people?

MR J. C. HANBURY

It works very well. But the answer to the last speaker is that it has to be planned, probably by the architect. The whole system has to be installed under control, and it is quite unsuitable for any attempt to adapt it to an existing conventional system. That is the answer: the conventional system and the system we have been talking about do not mix, and should not, under any circumstances, be allowed to mix. The whole thing must be tailor-made.

180

SIR FREDERICK WARNER, CHAIRMAN
This is the point at which we should break for tea and resume our discussion after tea. Immediately after tea I am going to ask Lord Simon to take the chair to excuse me to appear on your behalf at the "BBC downtown".

LORD SIMON, CHAIRMAN
We will now continue our discussion. Dr Holding.

DR J. A. HOLDING
Concerning Mr Drummond's figures on the ammonia concentration in the Trent, it is well known that ammonia can be nitrified to nitrate. I would like to ask Mr Drummond whether he has any data for the occurrence of nitrate in the Trent and whether he considers that this nitrate has originated from ammonia in the water or entered the watercourse as nitrate. Does Mr Drummond consider that some standards might be introduced to control pollution by nitrate and also phosphate?

MR I. DRUMMOND
I would remind you, ladies and gentlemen, that I am a lawyer, and I am not really certain what the questioner was after. Were you in fact asking about the concentration of nitrates in the Trent?

DR A. J. HOLDING
Yes, and whether this was related to any reduction in ammonia as it went down the river.

MR I. DRUMMOND
All I know of that is that at the moment the concentration of nitrates is not a problem, but, obviously, with the vast amount of sewage that we get in the Trent and the nitrification process going on, there will come a time when, clearly, we shall have to contemplate nitrogen-stripping plants. In other words, having gone right through to the line in your sewage works, and producing nitrates, you then have another piece of machinery on the end of that which strips the whole, or a large part, of the nitrogen out of the effluent. I suppose, though, that even then it could be a problem as far as the blue baby is concerned. As you know, any amount of nitrate, even small amounts in water, are dangerous; therefore, I suppose, if when we drink the Trent water, the baby ought to have a form of bottle water from another area. But we do point out that what we have in contemplation is a mixing of Trent water with other sources – relatively soft water from the top of the Peak District and other areas – so that, as a result of this, the nitrate concentration drops to fully acceptable limits. But I do agree with you that, perhaps towards the year 2000, this will be quite a severe problem. The other one, of course, is the problem of chlorides. Every time water goes through the human kidney, chlorides are produced. On top of that, of course, we have a number of other sources of chloride, including salt mines, brine-pumping in Stafford and, I suppose, at some time, when the degree of sewage has built up, also in the system itself. You know the kind of flat taste you get in London water, that comes mainly from chloride. At that point something will have to be done about it. I suppose the first thing to be done is to attack industrial

181

sources and to see if you can get those removed. After that, I presume there is the matter of partial de-mineralisation by whatever system will then be in use, and one certainly hopes that they will be very much better systems than are now available. You would set a nitrogen standard at the same time.

LORD SIMON
Could I ask just one question? I was under the impression that, if you have a river in the condition that I understand a large part of the Trent is, and certainly the condition that we have in parts of the Thames, in fact, in the first stage the sewage works after nitrification provide a reserve of oxygen. When the oxygen level falls low, the micro-organisms attack the nitrates and break them down.

MR I. DRUMMOND
That is perfectly true but I think that is the last reserve of nitrogen in the stream before you get starvation level. So, as far as we are concerned, what we want to achieve is a BOD of something like 6 as a maximum; 6 is the maximum recommended by the World Health Organization. Now, if we were to rely upon the oxygen in the nitrate, then I think the BOD would be very much higher than that anyway. It is a last reserve and, if that went, then you get bubbling H_2S coming off and you have a completely foul river which no-one can use at all. This, of course, is what we have in parts of our system now. This happens in the Irwell which, somebody said, reminds one of Stirwell.

PROFESSOR E. EISNER
Could I perhaps elaborate: Sir Frederick Warner was telling me about this yesterday. As far as I have been able to understand from any discussion here, the nitrates in themselves are not a demonstrable danger in any water. The algal bloom does not depend on nitrates but on phosphorus, because phosphorus is always limiting before nitrates are, as far as I have been able to see any evidence. The danger is the reduction of nitrates to nitrites, which can be a danger to small children. This reduction to nitrites will go on if, as a last resort, the micro-organisms have to feed on the nitrates. These nitrites then combine with ammonia, which is always present, making NH_4NO_2 ammonium nitrite, which equals $N_2 + 2H_2O$; the nitrite and ammonia just disappear as nitrogen and water, which, of course, is beautifully harmless.

MR G. W. HULL
The previous speaker said that conversion of nitrate to nitrite is not particularly harmful excepting in the case of methaemoglobinia in children; but nitrite is not the end of the story, because in the presence of some amines it can give rise to carcinogenic substances, which, if generated in the environment to any great extent and transmitted through the food chain, could lead to yet one more hazard of pollution. Very little attention has been given to the increase of potentially carcinogenic substances in the environment. Perhaps some of the medical or biological people here may wish to comment on this.

LORD SIMON
Is there anyone who would like to comment on that?

182

MR A. J. O'SULLIVAN
I understand your comment is mainly about carcinogenic substances. The immediate ones that spring to mind are certain hydrocarbons such as 3,4 benz-pyrene which has been detected in the Mediterranean both in man, animals and plants. I cannot think of any other specifically carcinogenic substances which are common as pollutants in man or in freshwater systems, but I should imagine that tar or its derivatives could contain such materials. Also there exists the danger either that new pollutants may have mutagenic effects or that such materials could be formed by chance reactions between several pollutants in the natural environment. Such a process need not be vastly different from the well-known production of photochemical smog by the action of ultra-violet light on pollutants from motor exhausts in Los Angeles.

MR G. W. HULL
May we not regard benzpyrene as merely a convenient generic index for a whole range of carcinogenic substances?

MR A. J. O'SULLIVAN
This is true; I believe there are a whole family of compounds, and as I mentioned a while ago they seem to have attracted most concern in the Mediterranean.

It occurs to me while I am speaking that the international aspects of sewage and industrial pollution do not seem to have received the attention at this session that they should be receiving according to the title of the conference. We have been confining ourselves very much to details of what happens in rivers, estuaries and perhaps coastal waters and I am wondering whether problems such as the present state of the Baltic, which is rapidly becoming very bad and suffering from almost complete de-oxygenation at its lowest levels, ought not to be a subject for discussion; and following from this perhaps we should discuss what happens internationally in such situations. The process of how scientists get together, information is exchanged and the formation of a working group such as the Swedo-Danish Committee on pollution of the sound between Sweden and Denmark ought also to be an important part of our discussion. The Swedo-Danish Committee carried out a very comprehensive survey of the effects of pollution in the area with which it was concerned and has laid the foundations for international co-operation and control. In the North Sea a similar system exists whereby information on pollution is exchanged between the countries bordering it; pollution of the Mediterranean is also the subject of a special international committee. Perhaps it might be instructive to consider where further areas of the sea are in danger or are in obvious need of international co-operation to halt increasing pollution problems.

DR J. A. G. TAYLOR (Unilever)
I would like to take up Mr O'Sullivan's plea for international consideration of the subject with an earlier question of what are multi-national companies doing about this. We are a multi-national company and, obviously, if you operate a factory in the United Kingdom, the problems that you are going to get from this are going to be very similar to operating a factory in Germany or any other part

of the world. I think we are trying to tackle this on a multi-national basis. Pollution can affect us in three ways:

(i) by virtue of the effluents produced by our factories;
(ii) in terms of pollutants that could be in the materials which we buy;
(iii) and then, finally, we have to consider the products that we manufacture and sell to the consumer, and which may have an effect on the environment.

To tackle these various aspects involves us in a very large and expensive effort.

To deal with factory effluents we have experts in various countries able to give advice to Unilever factories. These experts meet monthly to exchange information and to discuss mutual problems. To back this up we have a growing section, which I head, responsible for finding practical solutions to factory effluent problems. For this purpose we have assembled a multi-disciplinary group of chemical engineers, sanitary engineers, physicists, chemists, biologists and analysts. The first problem is to define the problems quantitatively and to cater for this need we are planning to establish a mobile laboratory.

What we have done is to centralise an operation which has operated in a more segregated fashion hitherto.

On the other side, of products, we are also extending and reorganising our quite considerable effort. There has always been a very considerable annual effort on toxicity testing and the group responsible has an international reputation for its work. Of course, the whole ecological question is so vast that one must draw upon as much expertise as possible. We have set up a standing committee of specialists from differing scientific disciplines, e.g. microbiologists, zoologists, biochemists, many of whom I know are known to the audience as internationally recognised experts. Again there is a practical arm responsible for running rigorous screening procedures to ensure that all new products are not harmful to sewage treatment processes or the environment.

A major problem in this area, as you have heard in the discussions these last two days, is that there are as yet no recognised criteria or test procedures adopted. This being so we are working with the international groupings in the various countries in which we are active in order to try and overcome this problem.

The concern over the potential contaminants in the raw materials we buy was covered in the comments I made earlier in the day.

Mr J. C. Hanbury
There are two points of a marine nature which I would like to make, and I had made a mental note to speak to this apropos Mr O'Sullivan's paper earlier on. First of all, industry regards the sea as, potentially, a very valuable recipient of industrial, and indeed other, effluents, subject to one's knowing precisely what one is doing in any particular place. At the moment, the whole trouble round our sea coasts is that we have hundreds of outfalls badly sited, totally inadequate in length – some of them not even going below low water mark – they are a scandal, it is a shocking situation. That, therefore, has created prejudice against the sea as a recipient at all, even if you know the character of your effluent, and the extent to which it has been processed. Secondly, you must know the precise tidal

pattern at the point of discharge; you must know what the prevailing currents are at that point. Wind is very important. You might not think so, but it is very important. The geology of the sea-bed, its steepness – whether it is shallow or whether it is deep – the type of vegetation on the sea-floor, supported by the geology of the sea-bed; if you study all these factors then the experts can say, with absolute safety, that if you put an outfall of such and such a length at such and such a point it can support an effluent of x, y, z characteristics and be absolutely safe and will do no harm at all. I hope there is somebody from the Water Pollution Research Laboratory here, since this is based to no small extent on years of work that they have done in the field and in the laboratory. I believe that if this matter is handled with great intelligence, with great circumspection, valuable use may be made of the sea, without prejudice to the amenities.

Dr J. A. G. Taylor
Have you seen the photographs from the NASA satellites using infra-red films and how this shows that the effluent does not mix?

Mr J. C. Hanbury
Yes, and that is vital; the photography from the air is part of the exercise.

Professor R. B. Clark
I cannot subscribe to the last comment. All effluents end up in the sea and with increasing emphasis on using rivers and lakes for drinking water rather than as open sewers, there will be an increased tendency for waste-disposal in the sea. Now, while we have some idea of the capacity of fresh waters to receive effluents we are unable at present to predict for the sea.

I agree entirely with Mr Hanbury that in a local situation it is possible to site a new effluent or sewer pipe in such a way as to distribute the effluent widely and do the least possible local damage, though I am not confident that the necessary preliminary studies to achieve this are always carried out. But in this situation it is essential to look a little further ahead.

It is already very well known that the world human population will double within the next thirty years and the population of Great Britain will undergo a very substantial increase. I find this very difficult to comprehend but the great increase in population will certainly come and it will change the whole situation. Thirty years is a very short time and it seems to me that we should be looking even beyond that when more radical solutions to problems of effluent disposal will certainly be required.

A few simple calculations show how serious the problem is. I have just seen one estimate for the east coast of the United States which shows that if the total estimated volume of effluent discharged at the present time is diluted to a safe and acceptable level, the total amount of sea-water required is not much different from the annual total exchange of sea-water on the continental shelf. Such calculations are only very approximate estimates and may be subject to a considerable error. But even if the error is of the order of ten times, it means that we are already within sight of reaching the capacity of the oceans to accept wastes from the highly industrialised urbanised population on the eastern seaboard of North America. The Baltic is certainly receiving more waste than it can safely absorb and the North Sea also may not be so far from being used to its capacity.

The sooner we realise that the sea has not an unlimited ability to receive wastes the better. We have no real idea how much longer we can continue as we have been doing, but with the rapidly expanding population, that time is almost certain to be much shorter than we suppose. It has even been proposed that instead of thinking about new sites for disposing of effluents we should really be thinking of avoiding producing effluents at all.

If one considers changes in policy and practice of effluent disposal that have occurred in the period 1920–70, it will be appreciated that fifty years is not a long time. We have less than that period before us in which much more radical changes must be made and time very definitely is not on our side.

PROFESSOR E. EISNER
I think, first of all, when I shook my head while Mr Hanbury was speaking, it was not because I did not agree there were places where one might conceivably discharge it, but because, as a scientist, I know that he is quite wrong in assuming that we are as infallible as he apparently does. On the contrary, I think we are very well aware of the fact that we cannot make most of the statements that he might think we can. One very, very simple example I think will show it more clearly than anything else, because it is perhaps the most fully studied of all cases that are relevant to this. Before the Windscale outfall was made, a very elaborate and expensive test was made to see how long it ought to be, and from this a curve was drawn of the logarithm of the concentration (C) of radioactive material on the shore-line versus the logarithm of the length of the pipe (L). Somebody later decided they were not satisfied with this and a second extremely expensive operation was mounted. In the range in question, one of these tests gave something like C is proportional to L^{-3} and the other one gave C is proportional to L^{-10}. Now it does not take very much knowledge of mathematics to see that this makes an enormous difference to what you plan to do. If you take the most conservative estimate anyone can give you, the only reaction you get is that no-one can afford to put the thing into practice. Remember that this is the most intensively investigated example there is; if you looked at any of the other cases you would find that the answers were very much less certain than they are in this case. If I could come to what Professor Clark was saying a moment ago, I think he is being pessimistic and I do not think there is any reason to overstate one's case – it just means that one gets less believed. I think what he has done is that he has assumed that every pollutant that goes into the effluent and into the sea is permanent (Professor Clark shakes his head). Well, in that case, what is the relevance of the figures you have given? You need to bring in, as I have said before – this is the third time and I hope it can be said again by other people – you need to bring in the decay time of the pollutant.

PROFESSOR R. B. CLARK
I was talking about per annum.

PROFESSOR E. EISNER
Yes, but you were assuming that all those pollutants that entered per year were

still there in that water at the end of the year and would have to be swept out to sea, no? Well, maybe not, that was not obvious.

Dr G. P. Hekstra

I was very much pleased by what Professor Clark said because in Holland we had for a long time the idea that, for instance, the Waddensee would be sufficiently cleaned by the water of the North Sea which enters it daily. We have recently discovered that the recovery of the Waddensee is not so quick because the North Sea seems to be more severely polluted and its water forms a real danger for the Waddensee.

Mr G. W. Hull

Many times during the last two days, participants in the Conference have referred to data which might be shortly summarised as biological indices of the quality of the living environment. Notably that concerning birds and plankton. The reduction in numbers of plankton may be due to seasonal variation, but some believe it is caused by oceanic pollution. The general principle arising is that somewhere there should be a collection of many more of these biological indices so that matters, such as these we have been considering during this Conference, could be related with them. In doing this, national economic plans should be borne in mind. An international institute on these lines may be needed, or perhaps better, more co-ordinated use of existing research centres which are already studying some aspects of these matters. I am sure that existing co-ordination in the collection and use of this data is inadequate.

Dr W. R. P. Bourne

If I could take up the Waddensee again. My understanding is that the big incident there was caused by something which I, personally, would never have suspected. It was a repeated leak from a pesticide factory on the lower reaches of the Rhine. The coastwise drift off the water went up the Dutch coast and into the Waddensee and poisoned very large numbers of birds in it, mainly Eider ducks and terns.

Dr G. P. Hekstra

In another instance unlawful release of a lot of copper-sulphate on the beach at Katwijk (near The Hague) poisoned the entire coastal North Sea and was ultimately found also at the entrance of the Waddensee. It caused great damage to many crustaceans and fishes.

Dr W. R. P. Bourne

Yes, the copper was another case. But I think it was the effluent from a pesticide factory that killed all the terns. It came from the mouth of the Rhine, up the coast and into the Waddensee and killed some 30,000 Sandwich terns, among other birds. It came from a completely unexpected direction, into a totally different inlet; personally, I would never have foreseen that happening. Another case where we are having trouble is with the breeding birds in the Shetlands and Faroes. The auks are producing the familiar picture of thin, breakable eggs, with too much pesticide in them for them to be fit for human consumption. The pesticide is presumably not coming from Europe at all, but from North America

187

in the North Atlantic drift, which shows the sort of distance you can get these effects. My feeling about this is that what we need now is some sort of estimate of the total world use of the dangerous substances, so that we at least have some idea of what is going into the sea and who is putting it there, to indicate where we shall have to apply controls some day, if it is going to be controlled. We need to know what the world production is and who is using it.

MR R. F. PEARSON
I would like to put a point here. You know the GLC have considered discharging sludge out to sea by pipeline. We did not know what the effect was going to be, so we thought the best way to go about it would be to try it out. In fact, we first of all discharged sludge at various points in the North Sea and found out the depths of water above the bottom. It came to the surface in a plume and the mass of the sludge settled out at about 15 ft below the surface. So we put in drogues at 15 ft down and did float tests as for the 15-ft drift, the surface drift and the bed drift. We then tried various points until we discovered one at a distance of $9\frac{1}{2}$ statute miles (8 nautical miles) due east of North Foreland where the drift did not come back to the English coast or into the estuary of the Thames. Having found that point, we then sent sludge vessels out, each carrying 2,000 tons, and each vessel discharged its 2,000 tons over a period of 12 hours through a pipe to the bottom of the sea, which at that point was 60 ft down. The sludge particulates were labelled with radioactive tracers, they were followed through and traced to where they went and analyses of the samples of the sea were taken and the polluting effect of the sludge on the sea was measured. I have got here a paper reporting the conclusions which were reached on this.

The first one was that the tidal action superimposed on a north-going residual current, combined with the low settling velocity of digested sludge particles, would lead to a general dispersal over the North Sea. Some deposition of the sludge would occur in the vicinity of the outfall, but this was not expected to bring about any measurable oxygen deficit in the overlying water column. Thirdly, there was no reason to suspect that there would be a measurable pollution increase on the nearby beaches, provided all floatables were removed from the sludge before discharge. Fourth, complete elimination of surface effects of the sludge discharged near the outfall was unlikely. Fifth, some entry into the Thames estuary of a part of the sludge discharge would occur, but it would be unlikely to represent more than a small fraction of the existing accretion load of the estuary, and at least twenty-eight days would elapse before it reached the beaches. Sixth, the pilot study discharge did not provide evidence of any changes on the sea-bed other than those which could be attributed to seasonal factors. Seventh, the chance of extensive accretion in the vicinity of the proposed outfall was remote. And eighth, and last, we concluded from this that it was feasible, from an engineering point of view, to lay a submarine pipeline to a point 8 nautical miles off the coast. Now, the sea itself has a purifying capacity, and we should be foolish, I think, as long as we do not poison it or put into it persistent poisons or impurities, not to take advantage of that capacity. After all, it is a natural asset and if we do not use it we have to use land, which would be much better used for other purposes. This test was scaled up – as I said, we discharged 4,000 tons over 24 hours for 3 months and tested it all the time – and the conclusions I have given here were ones that were arrived at by the WPRL and the HRS extrapolated to apply to the full quantity of sludge which we might, in about eighty years' time, build up to – that is to say, 10 million tons a year.

Dr J. W. Hopton
Can I add one or two things to support Professor Clarke's statements. Nowadays we know a great deal about the mathematical rules which govern microbial growth and metabolism. We can visualise a bacteria bed or an activated sludge tank in a sewage works as being essentially a continuously operating mixed microbial community. The community does not have an infinite capacity for assimilating the material it receives and if it receives too much the excess will go right through the sewage works into a river and eventually the sea. Similarly, the river and the sea have a limited capacity for assimilation. Thus if one reaches a situation where a biological system is being asked to exceed its theoretical capacity for degradation and assimilation, then waste material will accumulate. Nowadays the problem is aggravated by nutrient imbalance of products of modern technology which eventually find their way to the sewage works. The carbon : phosphorus ratio of most living materials is roughly the same and for this reason a sewage works is usually a very efficient purifier of waste which is of animal or vegetable origin. But if we now consider the carbon : phosphorus ratio of a household detergent this is very different from that of organic matter. The detergent powder contains far more phosphorus than the micro-organisms in the sewage works can assimilate, and inevitably the phosphate content of sewage works effluents has risen dramatically. Thus we can arrive at a situation, as Professor Clarke says, where the natural capacity of a system is overwhelmed by the character and quantity of the material which is applied.

Lord Simon
Could I just before the next question ask one arising out of that? You speak of the upset in the balance through detergents, but, surely, detergents are still, and will remain, a very small proportion of the total.

Dr J. W. Hopton
But the concentration of phosphate in the sea is generally exceedingly low. If you are increasing the local concentration, perhaps, one thousandfold, by even a small anount of detergent phosphate, this can easily upset the metabolism of the system.

Mr A. J. O'Sullivan
I would like if I may to make a few remarks following Mr Hanbury's comments. First of all I was delighted to hear him because I understand his background is that of an industrialist and his attitude towards pollution seemed to be a most responsible one. Though I believe that it is right to maintain the sea in as good a condition as possible and I work for an authority that seeks to do just this, I agree with Mr Hanbury that it is a legitimate use of the sea's resources to use them to purify or receive waste. However, if we are to do this in a responsible way there are a tremendous number of difficulties to be overcome. Mr Hanbury said we must know exactly what we are putting into the sea; yet from my own

experience much of the untreated effluent being discharged at present has a completely unknown composition. One cause of this I understand is that untreated waste is made up of process effluent from a number of different activities and reactions within the same or adjacent premises. The chemical engineers are interested only in the major product of the process reaction; the side reactions which constitute waste in many cases remain unknown in detail. Further mixing of untreated wastes either in containers or in vessels taking them out to sea for discharge makes the result even more unknown and lessens our chance of being able to forecast what happens to the effluent in the marine ecosystem.

I agree that tidal patterns, currents, winds, geology and the nature of the sea-bed are also very important and must be known; for example, surface water and anything on it such as oil moves at 3·4 per cent of the wind speed; material that sinks is subject to tidal and residual bottom currents and may build up on the sea-bed or be carried away, depending on the minimum velocity of water required to keep it in suspension or scour it off once it has been deposited. We also need to know very much more about basic marine ecology; about the interactions between the main species of plants and animals in both water and sediments, about the complex food webs in which they are linked. Processes of production and breakdown in the sea are known but before adding any organic pollution we would need to quantify such processes and be able to say how much pollution could safely be added before undesirable alterations of the ecosystem took place. Now when one considers the extent of this knowledge it is obvious that the gaining of it would be quite a stupendous task; the knowledge required far surpasses our present quantity of information about the sea and would require a considerable increase in the amount of effort at present devoted to marine research. So that while I agree that with such knowledge we could certainly design satisfactory outfall systems and use the sea's resources legitimately without causing harm to the environment, we cannot do so satisfactorily at present. Progress in our use of that particular resource of the sea must be slow and cautious.

There was also a mention by Dr Hekstra of a pollution incident on the Dutch coast involving a copper-containing waste. This incident was a very good example to show that the sea lacks the capacity for infinite dilution. The incident was due to the clandestine dumping of copper sulphate on a beach near Nordwijk on the Dutch coast in 1965. The resulting poisonous body of water on this open shore dispersed very slowly despite waves and tides, and while it moved 100 km to the north in fourteen days it remained close to shore and was diluted only five times. It caused extensive fish mortality.

Finally I would like to comment on the carbon:phosphorus ratio referred to by Dr Hopton. In a previous paper of mine I mentioned this point and suggested that we should be able to change the composition of sewage to suit an existing or desirable ecosystem. One way of doing this, first put forward by Stumm (1962),[1] is by making sure that our wastes contain the same relationship between

[1] Stumm, W. (1962). In discussion following "Methods for the Removal of Phosphorus and Nitrogen from Sewage Plant Effluents" by G. A. Rohlich, *Proc. 1st Int. Conf. Wat. Poll. Res.*, London, 1962, pp. 216–30.

190

carbon, nitrogen and phosphorus as does living matter; that is, if we put 106 parts of carbon, 10 parts of nitrogen and 1 part of phosphorus then the polluting effects of wastes would be minimised. Stumm suggested that most municipal sewages are nutritionally unbalanced in that they are deficient in organic carbon and that a significant fraction of the organic carbon in sewage cannot be assimilated. Seen in this way, conventional sewage treatment which mineralises substantial amounts of organic substances but which is not capable of eliminating more than 20 per cent to 50 per cent of nitrates and phosphates would seem to accentuate rather than diminish the problem. I do not know whether this idea has ever been tried in practice but I would be most interested to hear any observations or comments on it.

SIR FREDERICK WARNER, CHAIRMAN (returning)
May I apologise and thank you very much for taking over, sir.

I would like to come back to this question of what the environment can accept. I think what we must do is not to be discouraged but try to do some practical work – that is what we are here for as scientists and engineers. I think the encouragement which has been given by Lord Simon's Authority and the work that has been done by the Water Pollution Research Laboratory really should be taken into account. This is the result of one investigation on a semi-marine environment, in other words, of the tidal estuary of the River Thames. It will tell you a great deal of what the capacity of a particular system is to respond, how you can get a mathematical model which you can programme, and ask it questions, make predictions and deal quite well with what is going to happen in the future. Now, Mr Pearson will forgive me if I refer to something which took place last year when the Greater London Council had great difficulty with regard to a strike by the men who run their sludge boats. This resulted in a situation where there had to be a discharge of partially treated sewage and also sewage sludge to the Thames estuary. They dumped it, by agreement between the Greater London Council and the Port of London Authority, as an operation to minimise the effect on the environment and to study what happened – a most interesting scientific experiment. One of the things that I should point out to you is that when you discharge only partly treated sewage you put up the amount of ammonia, the ammonaical-nitrogen, in the river. This is what happened in the River Thames. Now there was, at the time, coming down from the Beckton Sewage Works, properly treated sewage which had had tertiary treatment in which the ammonia had been converted to nitrate. There was in the river a reserve of nitrate and when the slug of oxidisable material came into the river from the break-down, the nitrate came into operation as a reserve of oxygen once the dissolved oxygen was fully depleted. It was reduced to nitrite and, during one of the days of discharge, the total mass of nitrogen calculated in the river at 800 tons suddenly dropped to 600 tons because 200 tons had disappeared by the combination of ammonia and the nitrite, to form ammonium-nitrite which disappeared as elemental nitrogen. These are the sort of things you cannot foretell. Again, two days after this emergency began, the wind increased from force 3 to force 5 and, when this happens, as Technical Report No. 11 will tell you, the coefficient of mass-transfer, the diffusivity, which is 3 cm/sec under normal conditions, increases tenfold. The effect on the sewage was that "God blew with his winds, and they were scattered".

This is what happens very often, but most of the things which we predict we have to predict for the worst conditions. On top of these, there is very often a great additional effect like the wind effect. It affects directly mass-transfer, or the rate of solution of oxygen in the river, and I really plead for all of us to engage in more detailed scrutiny of our research rather than more speculation. I wonder whether we could get any evidence from biological indicators. You heard that in the tidal Thames, for example, there are now forty-two species of fish, whereas in 1956 there were none. Now, this is a biological indicator which I would accept. Something has happened; the fish have voted with their fins. We have got Professor Arthur here; he could probably tell us a great deal more because he has a research programme going on this. I think a good deal of patient work would fill gaps in our knowledge, and my hope is to thread our way through some of these difficulties.

LORD SIMON
I think you would be interested to know, Sir Frederick, that in your absence Mr Drummond has explained this position about nitrites and the freeing of ammonia. You confirmed precisely, from actual experience, what he told us would happen in certain circumstances. We were really, although these problems are closely interlocked, at the moment dealing with the capacity of the sea, and I do not know if there is anyone else who wants to add anything on this.

DR W. R. P. BOURNE
One thing one needs to remember about this disposal of sludge in the Thames; the stuff is dispersed all through the North Sea. If it is fertiliser that is put in, it will fertilise all the fisheries of the North Sea; if there is anything deleterious it may kill the larvae of all the herring, the plaice, the entire lot.

PROFESSOR D. R. ARTHUR
In carrying out the faunistic survey of the Thames we were looking at it initially from the point of view of establishing a base line of faunistic studies for future monitoring of pollution in the Thames. In general the fauna of the Thames can be treated in three groups, *viz.*, the benthic or bottom living animals, the plankton or floating organisms and the fish which we are trapping off West Thurrock. We are obtaining a wide range of fish species but numbers are large in respect of two or three species only. On the other hand, we are getting only a few species of tubificids in very large numbers in the mud along a 70-mile stretch of the river. In other words, ecological diversity is not great in the muds and ecological stability seems to be maintained here with relatively few species. The plankton was sampled from the GLC sludge vessel and from facilities in their laboratory we were able to obtain continuous automatic records of oxygen content and chlorinity of the water. In this way it was possible to correlate plankton occurrence in relation to abiotic parameters from Beckton to Barrow Deep, where the sludge was dumped. The results were interesting in that we found a horizontal zonation of planktonic species, principally copepods, which form part of the food of fishes.

192

My major complaint with regard to the ecology of polluted waters in Britain, and internationally as well, is the lack of co-ordination between authorities and people interested in pollution problems at the fundamental level of, for example, the comparative relationships of organisms in polluted and unpolluted waters. Closer co-ordination could (a) save ourselves considerable sums of money, (b) prevent us duplicating effort, and (c) working on a team basis from a large number of centres, would be very much more productive than it is at the moment. With regard to oil pollution, for example, Professor Clark has referred to the insidious effects of oils on fauna in the sea, but here again we are very largely ignorant of cause and effect, and frankly I am concerned with what might happen, too, if we had a spillage of the type of the *Torrey Canyon* tomorrow. How far would we still be trying to cure plague with aspirins as we did in 1967?

LORD SIMON

Well, I think we have reached the stage where we have, perhaps, got off the very important subject of whether we are overloading the sea. I wonder whether we could spend a little time on something which, indeed, as has just been suggested by Professor Clark, was touched upon by Sir Frederick and I think was brought up by Dr Howells, and that is whether we have adequate machinery for co-ordination and for the exchange of information, not merely nationally but internationally. Perhaps we could start nationally, because there are organisations within the country at work, and one would hope that they have some contact with each other. But, beyond that, what facilities are there? What can anyone do to help to improve the facilities for interchange of information and research internationally?

MR M. OWENS

If I may make one comment on this with regard to the work on enrichment or eutrophication; nationally the Fresh Water Sub-committee of the Natural Environment Research Council have set up a working group to review this problem in this country, to make recommendations with regard to the research which is required to be done, also the financing of the research and the extent of the effort. Internationally, the work on eutrophication has been organised through the auspices of OECD. Their commissions have consultants. One of their Water Management Groups set up a small working party on the problem of enrichment. They commissioned a consultant to review the whole field of nitrogen and phosphorus. They called together working groups of specialists who are working in the field to review that document and make suggestions and amendments to it. The document was then forwarded for further discussion. Subsequently they got the same group of specialists together to draw up a programme of what they thought were the priorities that ought to be tackled in the enrichment field, and this was referred back to the Water Management Research Group, who, I think, have taken cognisance of it – what they have actually done about it, I do not know – but, certainly, the attempt was made to organise research internationally, by harnessing international knowledge, particularly in the field of eutrofication.

MR A. J. O'SULLIVAN

I wonder if I may begin by mentioning co-operation on a fairly small scale, not even on a national level but on the regional level. I will use as my example local

193

Sea Fisheries Committees and mention very briefly how they are organised. There are eleven local Sea Fisheries Committees around England and Wales; Scotland is separate. The committee for whom I work, the Lancashire and Western Sea Fisheries Joint Committee, is similar to the other ten in most respects. There are 86 members, of which 43 are appointed by constituent local authorities, the other 43 members include river authority and dock and harbour authority representatives and the remainder are persons whom the Minister of Agriculture, Fisheries and Food considers to be acquainted with the needs and interests of the sea fishing industry. The Committee is financed by means of a precept on the rates of the constituent local authorities and thus in many ways we are similar to a river authority. However, there are a number of important differences. The members appointed by the Ministry of Agriculture, Fisheries and Food cover a very broad field and include people who are working as fishermen or are involved in the processing or distribution of fish and, I believe this to be very important, the Committee includes three members from local universities. Other Sea Fisheries Committees have a similar membership and also include scientists from universities or marine laboratories dealing with biological problems. The actual work of pollution control is delegated to a smaller sub-committee and this sub-committee includes the university members, several other members with direct knowledge of marine pollution problems, together with working fishermen and local authority members. This liaison with local authorities and universities has helped us very much; there has been a constant exchange of knowledge and information, and the Committee has benefited where projects in the field of pollution or fisheries have been accepted as suitable for M.Sc. or Ph.D. work by university departments. As far as I know river authorities do not possess any scientific liaison of this type.

DR G. HOWELLS

I think I should speak for just a moment about my own organisation and say that some of the work that has been going on in the constituent bodies of the NERC has indicated that marine productivity is changing, although I would not say that we knew anything about the causes of that change; whether they are, in fact, natural or unnatural causes. NERC is strengthening its activities in the field of marine pollution; already there are some laboratories financed by it and there is a considerable amount of university work that we intend to pursue still further. We are very much aware of the need for co-ordination, and we are looking into the ways by which this might be best achieved. As far as international contacts are concerned, there are two major organisations with which we deal. One is ICES, the International Commission for Exploration of the Sea, which was primarily set up to look at the fisheries problems. This is very successful – it consists of a group of scientists who talk to one another, and because so many other ecological factors are involved with the study of fisheries, they do not simply discuss commercial fish stocks – their interests are much wider. In addition, there is an Intergovernment Oceanic Commission (IOC) (UNESCO) which considers wider topics, and, more recently, there has been the formation of GESAMP (the Joint Group of Experts on Scientific Aspects of Marine

194

Pollution), and it is thought this group will be quite successful in co-ordinating the activities of different European countries.

PROFESSOR D. R. ARTHUR

I would like to take the opportunity of informing the Conference of the existence of UDASIS and certain other units now set up at Manchester University. It is a unit for research into pollution, not fundamental direct research. We are established by grants from the Social Science Research Council and the Science Research Council jointly. It is a multidisciplinary group – we have an economist, a lawyer, a chemical engineer (whom we have still to appoint), an ecologist and so on. Our job is only a two-year exploratory project; to try to predict what the trends in the future might be; secondly, to look at what steps are necessary, either through administrative measures or through legislation and enforcement, in order that more effective action might be taken in areas where there is pollution and where pollution might be increasing; and also to give some indication where further research is necessary. This is a two-year exploratory project; we are just embarking on it now and we would welcome any assistance, help, information or co-operation from any other research bodies that are working in this field.

PROFESSOR R. B. CLARK

It may be useful to describe the collaborative study of marine pollution that has developed on the north-east coast. In 1967 it was discovered that a number of different research groups were engaged, quite independently of one another, on research that related to pollution of the sea. These groups decided to co-ordinate their efforts and the whole became a part of the British contribution to the International Biological Programme. Further research groups have now joined us and the north-east programme now involves about twelve separate research groups in six universities, three marine laboratories, the Ministry of Agriculture, Fisheries and Food laboratory at Burnham-on-Crouch, industry (in the form of ICI at present) and local authorities in the area. The area under review extends from the Tay to the Humber. The co-ordination of this very heterogeneous collection of research teams has been assisted by financial support from the Royal Society, which enables us to hold regular co-ordinating meetings.

There is at present an increasing volume of research related to pollution of coastal waters in progress in universities. I agree entirely with Professor Arthur that a good deal of effort is being dissipated because of the lack of information about work actually in progress and because of the lack of a co-ordinated effort. We have been particularly lucky in the north-east because a number of groups were working in geographically closely related areas and co-ordination has proved easy to arrange. I believe similar co-ordination of activities in other parts of the coastline would be valuable and would benefit very much from encouragement and financial support for this purpose.

By far the most productive way of tackling these very complicated problems is to arrange a co-ordinated effort. This kind of research is very demanding of scientific manpower and therefore expensive. There is a considerable scientific resource in universities and it seems to me wise that we should enable it to make a full contribution. They are unable to do this effectively so long as they are left to work in small isolated groups.

195

I would just like to add that this is not, of course, purely a British situation, because I know that on Long Island Sound there has been very intensive study of the problem of effluent disposal into the sound, with a large number of university groups co-operating and collaborating in precisely the same way but, probably, more effectively than we have done so far. They have been studying this problem in just this co-ordinated manner for the last four or five years.

Professor D. R. Arthur
I think the major problem as far as Britain is concerned is the lack of communication between the various groups undertaking research on pollution. This arises in part, I suppose, from the sources financing research, but on a matter of this urgency there should be some sort of central organisation to correlate information. Under the present circumstances and in the absence of such an organisation, the first essential is to improve the communication channels between one group and another on matters of fundamental research, so that the problems to be tackled are known and the best lines of approach can be discussed. By this means we would, at least, avoid duplication of effort and make more profitable use of time.

Dr G. P. Hekstra
I would like to speak on international co-operation rather than on co-operation within the United Kingdom, which is scarcely my subject.

Lord Simon
As I see it, once there is a co-ordination movement in one country followed by a similar move in another country, then it is easier for those to get together. Perhaps we could come back to your point, Dr Hekstra, later.

Mr I. Drummond
I just wished to say a very short word on the position in England and Wales. First, the amount of money in research; if we are to double our water demand and thus double our effluents, and double our standards for effluent purification, then, obviously, our capital investment is going to go up from something like the present £160 million to £400 million or more on present standards, by the the year 2000. This is not related to doubling the population because, obviously, the population, certainly of England and Wales, does not double in that time; it is because of the higher standards demanded. So that sort of order of money would be involved. Then, clearly, to my mind, the vote for research nationally in this field must be at least doubled or trebled or even more. The research called for would obviously lie with, in the first place, the Water Pollution Research Laboratory, but all other institutions at present doing it would be involved and the work should be extended further. There is clearly value in independence in research in any field, and I think particularly in this one. Agreed, that money is sometimes dissipated because of a lack of co-ordination, but on the scale one is really talking about, in terms of money, I do not think this matters very much at the present time. Clearly, in the long term, one will have to co-ordinate research and it is no secret that the Water Resources Board would then become concerned with the quality as well as the quantity of the water. A central water authority might be set up. It would be for that authority

to co-ordinate research, control it and to have the funds to allocate wherever they are required, but, clearly, on a far greater scale than at present.

DR G. HOWELLS
Could I add something about national co-operation between the Research Councils? There has recently been a working party, which has representatives from all the five Research Councils, to deal specifically with their collaboration over pollution problems. A report is being prepared; this has now been to each of the Councils, and has been accepted, and we expect to be able to publish this document as a policy for pollution on behalf of the Research Councils later on this year.

MR A. J. O'SULLIVAN
Again may I add a very brief comment on co-operation among people doing scientific work, which Professor Clark has described for the north-east coast. This is most desirable, but the co-operation which I was trying to stress when I was describing Sea Fisheries Committees was not just between people doing scientific work but was a co-operation between scientists, between those administering pollution control legislation, between people formulating that legislation, and between those in the fishing industry whose activities and resources that legislation is designed to protect. This type of co-operation I would regard as being not on one level but between several levels.

PROFESSOR E. EISNER
May I ask whether anybody from Dr Martin Holdgate's office is here at this conference or was invited, because this seems to me the one obvious co-ordinating group – an office set up directly responsible to the Cabinet?

MISS M. M. SIBTHORP
Dr Holdgate was invited personally, but he unfortunately was unable to come, nor could he send a substitute. As you know, the future of the office is not yet clear.

DR J. A. G. TAYLOR
Can I answer this comment? I think this is really coming to the crux of what the problem is. We have been talking only about water pollution, but in fact there is an interaction between water pollution, air pollution and solid waste disposal. Under the last government, a very senior man, right at the top, was to be appointed to oversee the whole situation. We have, in this country, too many organisations trying to co-ordinate events; and we now even have one group trying to co-ordinate the co-ordinators. We have, moreover, the situation of the Water Pollution Research Laboratory dealing with water, and the Warren Spring Laboratory dealing with air pollution, sitting about a mile from each other, and I think, correct me if I am wrong, responsible to different bodies.

MR M. OWENS
Not at the moment. From 1 August we will be responsible to different bodies.

DR J. A. G. TAYLOR
If we look at the United States, we find I think that they have been looking into these problems at rather greater depth and now we have President Nixon

bringing together all these various activities, which are increasing almost every week, under one umbrella so that the situation can be seen as a whole. I think this is really what we should be talking about doing. First in the United Kingdom and then in conjunction with all the other countries.

Miss M. M. Sibthorp
Might I emphasise that if we are going to set up a government department with a ranking minister to look after pollution, in the first place you should see that he has the money, and in the second place you should see that he has adequate staff, because without both those the Ministry is merely something intended to reassure the public.

Mr M. Owens (Water Pollution Research Laboratory)
The position at present is that the Warren Spring Laboratory is responsible for studies of air pollution and of oil pollution in coastal waters, while the Water Pollution Research Laboratory is responsible for studies connected with water pollution; these include treatment processes for sewage and for effluents from industry, and the pollution of rivers, lakes, estuaries, and coastal waters. At the time of presentation of this paper, both laboratories were in the Ministry of Technology, but the Water Pollution Research Laboratory was to have been transferred to the Ministry of Housing and Local Government on 1 August 1970. This transfer was postponed and the Laboratory is at present (November 1970) in the Department of Trade and Industry. It is intended that it shall be transferred to the Department of the Environment in January 1971.

I would like to echo what Professor Arthur has just said. There is a need for a central co-ordinating body to deal with applications for research grants, because at present the same application is submitted to many funding organisations.

Lord Simon
I had rather hoped that the report that Dr Howells spoke to us about may help, at any rate, to concentrate our minds on the necessity of pulling this thing together. Now, I did promise Dr Hekstra that he would come back and talk about co-ordination with other countries.

Dr G. P. Hekstra
At the beginning of the session yesterday morning there were some introductory remarks on international co-operation, and I would specially draw attention to the work which is already going on, on an international basis, within the framework of the International Biological Programme (IBP). As you may all know, IBP will end in 1972 and then the work will be continued in the new UNESCO programme which will go under the name of "Man and the Biosphere". You can already see from this name that it has a wider scope than biology alone. It is not only an extension of the International Biological Programme, but it will deal also with the interaction of man with the biosphere, and the influence of the biosphere on man. In November last year (1969) the UNESCO working groups had set up the first drafts for this "Man and the Biosphere" programme, which, probably even now, may be in circulation within the international UNESCO delegations. I know that this will offer opportunities for very good

198

international co-operation in tackling the problems of pollution, especially through research. Some of the thirty themes of "Man and the Biosphere" deal especially with monitoring problems and with pollution research, and I hope that, within this framework, specialists will meet and discuss these problems.

The original IBP concept was mainly centred on productivity; there were some sections on terrestrial productivity, fresh-water productivity, marine productivity, a section on conservation, a rather small section on problems of use and management and questions of pest control, particularly biological control of pests, and there was a very large section on human adaptability problems. The Dutch UNESCO delegation for the coming General Assembly, which will meet in October next, is considering a proposal to give much more support to the UNESCO Department of Environmental Sciences, because we feel that its international funds to start a really good and challenging co-operation in "Man and the Biosphere" need a very great deal more money than was provided. I will give you some figures as an impression. The whole section of "Natural Sciences and Their Application to Development" has a budget of about $45 million and only 22 per cent of the entire section budget is given to environmental sciences and natural resources research. Of this 22 per cent, approximately 19 per cent is devoted to integrated natural resources research and ecology, under which the "Man and the Biosphere" programme should be run, 34 per cent will go to the earth sciences, 17 per cent to hydrology and nearly 14 per cent to oceanography. So we, in the Dutch delegation, intend to make a proposal to increase the budget devoted to integrated natural resources research, because we feel that if we cannot get a sufficient international funding, we can never achieve good international co-operation, since co-operation costs money (for meetings, courses and training programmes). I would be very pleased if there was someone here from the British delegation to the UNESCO Conference, who would care to see our report on this. There could possibly be some concerted action of the Dutch and British delegations.

LORD SIMON
Is there anyone here who could pick up that invitation?

DR G. HOWELLS
I have a colleague who is responsible to international organisations; I would be happy to receive it for him.

SIR FREDERICK WARNER
Can I strike a discordant note?

LORD SIMON
With the greatest of pleasure.

SIR FREDERICK WARNER
As it happens there is an argument against international co-operation seen from the point of view of an engineer. I talk to my engineering colleagues throughout Europe and we regularly discuss the problems which are facing us. One of our problems is that we are always under criticism from university scientists, and yet at the same time have to keep the whole junk-pile continuously running. Now,

199

this is not an easy job at all. It is alright to criticise, but I am afraid that engineers on the whole find themselves in the dock and they say, "Yes, we have done it. We have taken action, because action had to be taken." You find sometimes, and I become nationalistic on this point, that we are subjected to heavy criticism, say from Holland. I would like to try to put into perspective what *we* do to the North Sea compared with what Holland does to it. The Thames, for much of the year, puts out, as we have heard, a minimum flow over Teddington Weir of fresh water of 170 million gallons a day, say 9 cubic metres a second. Now, Rotterdam has a scheme to discharge excess water through the Delta Barrage when the flow through Rotterdam exceeds 6,000 cubic metres per second. If you want to look at the various contributions to the North Sea, you ought to see what is discharged through the Rhine. We have not anything in this country that compares with this sort of thing. The average flow of the River Thames is only 70 cubic metres per second throughout the year; the average flow of the Rhine is something in the order of 1,500 cubic metres per second. We are not operating on the same scale. You have to watch very carefully that you do not allow yourselves, as scientists, to become the tools of nationalist arguments because it is possible for this to happen. As a Dutchman you can argue that, because you are lower down the stream, you are therefore "pure white"; that the French potash industry in Alsace discharges two millions a year of common salt into the Rhine and you have to receive this in Holland. But the whole of the petro-chemical industry in Holland discharges whatever it wishes directly into the same river from Rotterdam, and it goes straight into the North Sea. For a long time to come we must ask, when talking together in private meetings, that engineers who are trying to work with applied scientists should have a better understanding with their academic colleagues. We have got to try and talk frankly, because it is inevitable that commercial positions will be taken by competitors. National sovereignty in each country would demand that we make what is the best use of what we see as a national resource, and in England a national resource is a long coastline, coupled with very short rivers. We do not add a great deal to the maritime environment, though large continental countries do. We would like to know what they intend to do about it, and must use our coastline just as much as Norway and Switzerland use their hydro-electric potential, as something which is a property of their country. So, if this may seem to strike a discordant note, I think it is one that we have to take into account, because most of us earn a living by keeping wheels turning.

MR F. MACDONALD
I think this highlights the problem, and I am sure that any layman who has been listening to what has been going on during the last two days must be absolutely staggered. All this information; all this knowledge, all these people working at this problem. We know what the pollutants are, we know what are degradable, we know what are non-degradable. Surely it would be possible to do an inter-national materials balance. We know the sources of all these products, where they are going to, but we are getting our priorities wrong. We must have an international organisation which will develop a materials balance for all these products. Here we are, with all this computerisation – we can send a man,

within 50 ft, somewhere into another world – but we cannot trace where some of these things are going to around our coast and into our oceans. It really is a stupefying situation.

LORD SIMON

If we are having any discordant notes thrown about, perhaps I can throw one in. I must admit, listening to the earlier part of this discussion about the sea getting full of what we throw into it – we really could not put in any more – I felt rather like Old Bill in Bruce Bairnsfather's cartoons: "If you know a better 'ole, go to it." I mean, it may be absolutely disastrous to put all this stuff into the sea, but where else are you going to put it? Somebody made what I think is an important suggestion, that perhaps what we have really got to look at in years to come is how not to have any waste. I confess that I do not understand where that is going to lead us; there must always be, I would have thought, some things, some products of industry, that are not wanted, unless we are going to have spoil-heaps all over the place, which, after all, none of us wants. But, at any rate, that may be an angle from which we ought to be looking at things. Not how do we get rid of this waste, but how do we reorganise our processes so that we do not have it? We know, however, of some natural processes where that can not possibly be done.

DR J. A. G. TAYLOR

Can I add a rider? There are also the possibilities of the beneficial use of some of these waste products. We talked about thermal waste, we know that fish can grow rather faster in warmer conditions. Fish farming is another project that is being looked at. Tie the two together, and, perhaps, with some of the organic pollutants that have got to be discharged, and you may have a viable proposition. I do not know how much has been done on this, but, obviously, there is a general commercial interest. I think this is something to think about, there are positive aspects to the problems.

DR W. G. MARLEY

I listened with interest to Mr Drummond making his plea for increased research on the basis of the huge capital outlay which will be necessary in developing the resources by the turn of the century, and much of the rest of the discussion has been about co-ordination and circulation of information. I would like to come back to one point. Experience with radioactive materials round the coast has shown that the problems in different places, with regard to these materials, are completely different. Whilst you may have one radio-biological problem in one location, it does not help very much in solving the problem in another location. So, in advocating extension of research, I think we must realise that it may mean setting up several research teams. Research teams have to be viable, and there have to be sufficient people working together, since isolated researchers very often do not do as well as teams. I think this underlines the scale on which there should be an increase in research in this field.

PROFESSOR A. NEWELL

I am probably the epitome of all the laymen who are here. Actually, I am a member of the Executive Committee of the David Davies Memorial Institute of

International Studies – neither scientist, nor administrator, nor lawyer. I have been very much puzzled by the terminology that has been used and by the diagrams on the board. I have been more intrigued with the spirit of co-operation that has been prevalent during these two days, beginning with the collaboration between the David Davies Institute and the Department of International Politics, headed by Professor Trevor Evans. There we had one illustration of how two people, two organisations, can come together and produce this important conference on pollution, and I am simply amazed at the wealth of knowledge that has been shown by so many here. This last session, it seems to me, has tried to tie it all together, and, though nobody has suggested that there is an ironclad method of doing this even nationally, let alone internationally, I wonder if this is, perhaps, a harbinger of the kind of thing we are aiming at. In private discussion here I have heard the comment, "What is going to come out of all this? We talk, but is there any action?" Well, I have not heard any specific programme of action. We now disperse to our various universities, institutions, research bodies . . . I do not know the answer but I like the spirit everyone has shown. I have rarely been to a conference where I saw groups of people talking so avidly together, obviously talking shop. Perhaps that was the purpose in bringing us together, to talk shop, each one's shop with the other's shop. And I just wanted to register my excitement that this has happened, and I hope – and it is only a hope – that something that has been suggested by one or another of us may come to some kind of fruition. This is one layman's judgment on the Conference, his hope that it will not be entirely fruitless.

LORD SIMON
Thank you very much, Professor Newell. I think that very aptly brings our consultations to an end, because, indeed, it is the time set for an end to our discussion. I am sure you would all wish at the end of this session to thank the speakers who introduced this last discussion, to thank, indeed, everyone who has taken part, and especially to thank the David Davies Institute and the University of Wales for their co-operation in making this possible.

Vote of thanks

MISS M. M. SIBTHORP
My Lords, ladies and gentlemen,
    In prefacing these few remarks, may I say that I would challenge Professor Newell's qualification of himself as the most absolute layman here, for I am myself the quintessential layman.
    In closing this Conference I feel that, especially in view of very much that we have heard, it may not be out of order to emphasise again that although awareness of the extent of the incipient and more insidious dangers of pollution of the environment is growing and becoming widespread, there is no room for complacency. We are faced, as has been acknowledged, with very considerable problems.
    "Efforts to improve the environment in Britain", according to the Chief Public Health Inspector of Warrington in an article in *Municipal Engineering*, "are at best mediocre and at worst criminally negligent." He goes on to say that

202

conservationists are "aghast at the self-congratulatory pronouncements of politicians and officials that much progress has been made, particularly in the field of air pollution" in general. And, as I said this morning, the Report of the Standing Technical Committee of the Ministry of Housing and Local Government is concerned at the recent increase of foaming on the rivers due to the rising use of hard detergents.

I think one of the chief points emerging from our discussions is the obvious need for much closer collaboration between the various disciplines involved, lawyers and politicians. The problem is, as our Chairman said in his opening address, and as has been reiterated since, essentially a political one, but political decisions should be based upon sound knowledge. There is need for a much greater interchange of information, both as to what is being done and what is needed to be done. What, in fact, the most urgent problems are. Moreover, it is no use blaming the public for lapping up sensationalism if that is about all they are offered.

I think, also, two further points should be borne in mind when discussing pollution. One is that statistics are the most flexible instrument ever bestowed upon a naturally devious race. And the second is that in any case they bear very little resemblance to what we see, feel or smell around us. If we now have to pay for what was once our heritage, clean air and water and uncontaminated land, we should be told how much and why, not simply ignored when the decisions are made.

Incidentally, I was somewhat startled to be told that the only living things which had any right to survival were those that were useful to man. But "Whom the Gods would destroy they first make mad", and who am I to question their methods?

It only remains for me, on behalf of the Institute, to thank all our speakers for contributing so greatly to the success of this Conference and to our Chairmen who have so skilfully steered our discussions. We are especially grateful to Lord Hodson who, at considerable personal inconvenience, has devoted so much of his time to our proceedings.

I am sure that we are all very conscious of how much we owe to the delightful hospitality extended to us by the University College of Wales and to the personal kindness of Professor Trefor Evans.

We thank all who have come here and hope that they consider that their journey has been worth while.

203

President Nixon's Message on Ocean Dumping, October 1970

To the Congress of the United States:

The oceans, covering nearly three-quarters of the world's surface, are critical to maintaining our environment, for they contribute to the basic oxygen–carbon dioxide balance upon which human and animal life depends. Yet man does not treat the oceans well. He has assumed that their capacity to absorb wastes is infinite, and evidence is now accumulating on the damage that he has caused. Pollution is now visible even on the high seas – long believed beyond the reach of man's harmful influence. In recent months, worldwide concern has been expressed about the dangers of dumping toxic wastes in the oceans.

In view of the serious threat of ocean pollution, I am today transmitting to the Congress a study I requested from the Council on Environmental Quality. This study concludes that:

the current level of ocean dumping is creating serious environmental damage in some areas;

the volume of wastes dumped in the ocean is increasing rapidly;

a vast new influx of wastes is likely to occur as municipalities and industries turn to the oceans as a convenient sink for their wastes;

trends indicate that ocean disposal could become a major, nationwide environmental problem;

unless we begin now to develop alternative methods of disposing of these wastes, institutional and economic obstacles will make it extremely difficult to control ocean dumping in the future;

the nation must act now to prevent the problem from reaching unmanageable proportions.

The study recommends legislation to ban the unregulated dumping of all materials in the oceans and to prevent or rigorously limit the dumping of harmful materials. The recommended legislation would call for permits by the Administrator of the Environmental Protection Agency for the transportation and dumping of all materials in the oceans and in the Great Lakes.

I endorse the Council's recommendations and will submit specific legislative proposals to implement them to the next Congress. These recommendations will supplement legislation my Administration submitted to the Congress in November, 1969 to provide comprehensive management by the States of the land and waters of the coastal zone and in April, 1970 to control dumping of dredge spoil in the Great Lakes.

The program proposed by the Council is based on the premise that we should take action before the problem of ocean dumping becomes acute. To date, most of our energies have been spent cleaning up mistakes of the past. We have failed to recognize problems and to take corrective action before they became serious. The resulting signs of environmental decay are all around us, and remedial actions heavily tax our resources and energies.

The legislation recommended would be one of the first new authorities for the Environmental Protection Agency. I believe it is fitting that in this recommended legislation, we will be acting – rather than reacting – to prevent pollution before it begins to destroy the waters that are so critical to all living things.

RICHARD NIXON

The White House
October 7, 1970

Speech of Hon. Mitchell Sharp, Canadian Secretary of State for External Affairs, introducing the Arctic Waters Pollution Bill, in the Canadian House of Commons, Ottawa, April 1970

I am sure that there is general agreement on all sides of this House with the two fundamental objectives underlying the Arctic Waters Pollution Prevention Bill. These objectives are the economic development of the Canadian Arctic and the preservation of a unique environment comprising land and ice and open sea. The Government has given long and careful study to the means by which these objectives could best be given effect and translated into legislative terms. We have considered these questions in the light of the duty and responsibility which Canada owes not only to itself but to the community of nations – that is to say, to mankind as a whole. We have refused to be stampeded by clamor from any quarter. We have rejected simplistic solutions which could create more problems than they might resolve. Instead we have evolved, after very wide-ranging deliberations, a constructive and functional approach which distinguishes between jurisdiction and sovereignty and between essential national objectives and chauvinism, which reconciles national interest and international responsibility, and which will prevent pollution without discouraging development.

The problem of environmental preservation transcends traditional concepts of sovereignty and requires an imaginative new approach oriented towards future generations of men and the plant and animal life on which their existence and the quality of that existence will depend. The problem of environmental preservation, moreover, must be resolved on the basis of the objective considerations of today rather than the historical accidents or territorial imperatives of yesterday. Canada has always regarded the waters between the islands of the Arctic Archipelago as being Canadian waters: the present Government maintains that position. The Bill we have put forward aims to meet a real and imminent problem. This exercise of jurisdiction for the purposes of pollution control can in no way be construed to be inconsistent with a claim of sovereignty. Similarly, the exercise of sovereignty over an area of the sea extending 12 miles from shore (in accordance with a provision embodied in another Bill) cannot be said to be inconsistent with a claim to sovereignty beyond 12 miles. There is excellent authority for these two related propositions: in the 1910 North Atlantic Coast Fisheries case between Britain and the USA, the Permanent Court of Arbitration held that a State may, without prejudice to its claim to sovereignty over the whole of a particular area of the sea, exercise only so much of its sovereign powers over such part of that area as may be necessary for immediate

207

purposes. That case is of particular relevance to the Canadian situation since it involved areas off Newfoundland, Labrador and other parts of Canada's Atlantic coasts.

There are those that argue that the problem of marine pollution can only be met by multilateral rather than unilateral action. Canada has attempted the multilateral approach to this problem, most recently at an international legal conference in Brussels in 1969. On that occasion, however, we were unsuccessful in our attempts to persuade the major shipping and cargo-owning States to provide adequate recognition and protection for the rights and interests of coastal States which are the innocent victims of pollution incidents on the seas. State practice, or in other words unilateral action by States, has always been a legitimate means open to States to develop customary international law. This is how the Three-mile Territorial Sea, and later the 12-mile Territorial Sea, originated. It was unilateral action by the USA in the 1945 Truman Proclamation which led to establishment of the Continental Shelf Doctrine in international law. It was the practice of Norway in connection with the delimitation of its territorial waters which introduced the straight baseline system later written into the Geneva Convention on the Territorial Sea. Again, it was by unilateral action that Canada in 1964 and USA in 1966 established nine-mile contiguous fishing zones.

The action we are proposing for the Arctic waters in no way rules out the possibility of developing international arrangements for the preservation of the marine environment in Arctic regions. The Bill we have introduced should be regarded as a stepping stone towards the elaboration of an international legal order which will protect and preserve this planet Earth for the better use and greater enjoyment of all mankind. A single ecological system governs the lives of all men, and the Arctic regions are an extremely important part of that system. They determine the livability of the whole of the northern hemisphere. This Bill is a beginning. It puts forward a legislative framework within which we will develop controls and safety standards to ensure that this unspoiled and uniquely vulnerable region is preserved from degradation. *We will consult with other countries before we promulgate regulations to this end.* We hope that these other countries will show a spirit of understanding and co-operation so that together we can construct a system of internationally agreed rules and safety standards which will advance our common interests without interfering unreasonably with particular interests.

Canada has a long tradition of leadership and active participation in multilateral efforts to resolve problems which go beyond purely national concerns. This is especially true in the field of international environmental law. In the famous Trail Smelter Case we went to arbitration with the USA in 1935 and accepted state responsibility for the pollution of USA territory. In later years we pressed hard for the Non-Proliferation Treaty and were in the vanguard of attempts to prevent fallout pollution from atomic testing. We have been engaged with the USA since 1909 in a unique experiment in international co-operation on common environmental problems, through the International Joint Commission. However, it is precisely this long experience with multilateral and bilateral approaches which convinces us that immediate action by Canada is required for protection of the Arctic environment. We know only too well that a situation requiring urgent action cannot be met by the slow and difficult process of negotiating international arrangements. However valuable may be the work of the International Joint Commission, citizens of both Canada and the USA are painfully aware that it has not prevented the pollution and contamination of the Great Lakes to the point where the very life of these vast bodies of water is threatened. The International Joint Commission is undertaking remedial action on the Great Lakes but that action is long overdue and will not easily undo the ravages that have taken place. We cannot be too late everywhere. We cannot wait until the damage has been done in the Arctic if only because such damage in that environment may well be irreversible.

The first attempts to find an international solution to the problem of pollution of the seas by oil were made in the early 1920s but did not achieve even partial success until the late 1950s. In 1926 an international conference held in Washington, D.C. drew up a relatively modest proposal for the control of deliberate marine discharges of oil or oily mixtures. Even this modest proposal failed to achieve ratification. By 1954 the oil pollution problem had reached such a state of crisis in some areas that a second major conference was convened. The result was the London Convention for the Prevention of Pollution of the Sea by Oil. This Convention, like the 1926 proposal, deals with the prevention of deliberate pollution by tanker cleaning operations, but leaves enforcement to the Flag States rather than the Coastal States suffering the damage. This Convention was adopted despite strong opposition from the USA, which believed that the problem of deliberate discharge would disappear by educational programs and technological advances.

The London Convention was only slowly accepted, and it was not until four years later that sufficient countries had ratified it to bring it into force. Canada's Instrument of Acceptance was deposited in 1956, and that of the USA in 1961. The Convention was amended by a second conference held in 1962 under the auspices of the Intergovernmental Maritime Consultative Organization. The 1962 amendments were relatively marginal but extended from 50 to 100 miles the minimum zones in which the deliberate discharge of oil is prohibited. Canada accepted these amendments in 1963, but they did not achieve sufficient acceptance to come into effect until 1967.

The amended London Convention remains the major international instrument in force in this field. Despite its modest aims, and despite the fact that it leaves enforcement to the Flag States and thus preserves their traditional exclusive jurisdiction over their vessels on the high seas, this limited Convention did not come into effect until some 30 years after the oil pollution problem first began to attract serious international attention. Its inadequacies as to the scope and enforcement of its provisions are, I believe, disputed by no one. More recently, however, we believed there was cause to hope that the nations of the world might join together to attack the problem of oil pollution on a broader front and to adopt more effective measures for its prevention and control. The *Torrey Canyon* incident had awakened States and public opinion to the catastrophic consequences of a spill from a jumbo tanker. Domestically and internationally there had been increasing signs that the quality of the environment was becoming the major issue of our time. Against that background Canada went to Brussels in November 1969 to participate in an International Legal Conference on Marine Pollution Damage. The results of the Conference, however, while reflecting a certain degree of progress, were seriously disappointing.

Many delegates at Brussels displayed what to us appeared to be an excessive caution and conservatism and a rigid preoccupation with the traditional concept of unqualified freedom of the high seas. That freedom in our eyes seemed to be tantamount to a license to pollute; it did not in any way strike a proper balance between the interest of the Flag State in unfettered rights of navigation and the fundamental interest of the Coastal State in the integrity of its shores. As a result, despite our most vigorous efforts, Canada was only partially successful in achieving recognition of the paramount need for environmental preservation and the principle that the bulk carriage of oil and other pollutants by sea is an ultra-hazardous activity which gives rise to an absolute liability to compensate in full the victims of pollution damage arising from such carriage.

The outcome of the Brussels Conference was so little oriented towards environmental preservation and so much oriented towards the interests of ship- and cargo-owning States that Canada abstained from voting on the Public Law Convention Dealing with the Right of Intervention on the High Seas and voted against the Private Law Convention on Civil Liability for Pollution Damage. While the main thrust of the Bill under debate is preventive, that of the Brussels Conventions is remedial and liability-oriented. I do not wish, however, to be excessively severe or negative in judging the achievements of the Brussels Conference. The Public Law Convention negotiated there incorporates the very important principle that Coastal States may intervene against foreign ships on the high seas to prevent or minimize major pollution damage where a marine accident threatening or actually causing oil pollution has already occurred. I must say in this connection that I find it anomalous that certain countries can accept the right of a Coastal State to sink a foreign ship on the high seas when a marine accident threatens pollution, but at the same time assert that Coastal States do not have the right to prevent such an accident by turning away such a ship from areas off its coast or by imposing certain safety standards or preconditions for entry into these areas.

The Coastal State's right of intervention on the high seas, as incorporated in the Brussels Convention on Public International Law, may perhaps represent a sufficient basis, for the time being at least, to protect the marine environment and Canada's coastal interests beyond the proposed 12-mile limit for our territorial sea on the Atlantic and Pacific. The problem of pollution in those areas is also a matter of vital concern and will be given the most energetic attention by this Government. With respect to the Arctic other measures impose themselves.

We hope that the Arctic Waters Bill will provide a framework for internationally agreed safety standards. The brief review of multilateral efforts which I have just made is sufficient proof, however, that an approach of that kind would not have met the urgent need for early action and would not have provided the stability and certainty required for investment in the development of Arctic resources and Arctic navigation.

There can be no doubt that Canada has tested the climate for international

210

action against marine pollution, and there can equally be no doubt that the climate has been found seriously wanting. We are determined to discharge our own responsibilities for the protection of our territory. We are equally determined to act as pioneers in pushing back the frontiers of international law so that the *laissez-faire* regime of the high seas will no longer prevent effective action to deal with a pollution threat of such a magnitude that even the vast seas and oceans of the world may not be able to absorb, dissolve or wash away the discharges deliberately or accidentally poured into them. The Arctic Waters Bill represents a constructive and functional approach to environmental preservation. It asserts only the limited jurisdiction required to achieve a specific and vital purpose. It separates a limited pollution control jurisdiction from the total bundle of jurisdictions which together constitute sovereignty. In this it resembles in some degree the approach which Canada was among the first to adopt with respect to jurisdiction over the exploitation and conservation of fisheries resources. The results which have been achieved in the latter field encourage us now to lead the way in developing rules to prevent pollution of the sea and of the shores of Coastal States. We firmly believe that this is the best way to bring order out of impending chaos in the law of the sea.

The pioneering venture upon which we are embarked is a measure of our serious concern at the failure of international law to keep pace with technology, to adapt itself to special situations, and in particular to recognize the right of a Coastal State to protect itself against the dangers of marine pollution. Existing international law is either inadequate or non-existent in this respect. Such law as does exist, as I have already indicated, is largely based on the principle of freedom of navigation and is designed to protect the interests of States directly or indirectly involved with the maritime carriage of oil and other hazardous cargoes. A new "victim-oriented" law must be created to protect the marine environment and those rights and interests of the Coastal State which are endangered by the threat to that environment. The Arctic Waters Bill is intended to advance the development of such new law. It is based on the fundamental principle of self-defence and constitutes state practice, which has always been accepted as one of the ways of developing international law.

Where the law is deficient any action undertaken to remedy its deficiencies cannot properly be judged by the existing standards of that law. Such a proceeding would effectively block any possibility of reform. Canada remains firmly attached to the rule of law in international affairs and has the highest respect for the International Court of Justice and the part it plays in the maintenance of that rule of law. At the same time, however, we are not prepared to litigate with other States on vital issues concerning which the law is either inadequate, non-existent or irrelevant to the kind of situation Canada faces, as is the case in the Arctic. It is no service to the Court or to the development of international law to attempt to resolve by adjudication questions on which the law does not provide a firm basis for decision. For these reasons we have been obliged to submit a limited new reservation to our acceptance of the compulsory jurisdiction of the International Court of Justice.

Even with the new reservation, Canada's acceptance of the compulsory jurisdiction is much broader than that of many other countries. It does not in

211

any way reflect lack of confidence in the Court but takes into account the limitations within which the Court must operate and the deficiencies of the law which it must interpret and apply. Moreover, it may be revoked and Canada's acceptance of the compulsory jurisdiction may again be broadened at such time as those deficiencies are made good. In the interval Canada stands prepared to appear before the Court where the Court is in a position to exercise its proper function and render a decision either for or against us. Such is the case, for instance, with respect to our Bill on the 12-mile territorial sea. Our readiness to submit to the international judicial process remains general in scope and is subject only to certain limited and clearly defined exceptions rather than to a general exception which can be defined at will so as to include any particular matter.

I have already stressed the Government's hope that it will be possible to achieve internationally agreed rules for Arctic navigation within the framework of our proposed legislation. We recognize that the interests of other States are inevitably affected in any exercise of jurisdiction over areas of the sea. We have taken these interests into account in drafting our legislation; we have, for instance, provided that naval vessels and other ships owned by foreign Governments may be exempted from the application of Canadian anti-pollution regulations if the ships in question substantially meet our standards. We will give the interests of other States further consideration by entering into consultations with them before promulgating safety regulations under the Arctic Waters Bill.

I should point out that the interests of other States in the uses of the sea are not necessarily in conflict with ours. We too are concerned to preserve the essential freedoms of the seas. We too do not wish to place unnecessary or unreasonable restrictions on maritime commerce. Security factors are vital to us as well as to others. It is because we share the concern to head off developments undesirable for common interests that we ask other States to adopt a flexible attitude which is responsive to new needs and special circumstances, and that we seek the co-operation of other States and offer them ours.

In recognition of common interests and in the spirit of co-operation Canada has for many years engaged in periodic consultations with the USA on matters concerning the law of the sea. We have not always agreed on those matters but we have always benefited from obtaining a better understanding of our respective positions and concerns. I would like now to turn to a point of some importance in considering the international aspects of this legislation, namely the position of the US Government concerning it. The Government of the USA has on a number of occasions recently expressed a particular interest in the various aspects of the law of the sea raised by the Prime Minister's Statement in the Throne Speech debate when he announced the Government's intention to introduce legislation to protect the ecological balance of the Canadian Arctic, and requested an opportunity to discuss them with us. Two rounds of discussions were held for this purpose. On 11 March the Canadian Ambassador to the USA, Marcel Cadieux, accompanied by two Canadian officials, Mr Beesley, Head of the Legal Division of the Department of the Exterior, and Mr Head, Legislative Assistant to the Prime Minister, called on Mr Alexis Johnson, USA Under-Secretary of State for Political Affairs, and a group of senior USA officials. These discussions were very frank and friendly but they revealed, as expected, differences of views between our two Governments on a number of questions, and it was agreed that a further round would be held after the USA Government had had time to consider the matter further. On 17 March President Nixon phoned the Prime Minister to express his interest in the matter and offered to send a high-level team to Ottawa for further discussions. On 20 March a team of senior USA officials led by Under-Secretary of State Johnson and including the Under-Secretary of the Navy and an Assistant Secretary of Transport, as well as senior officials from the State Department, Defence Department, the Coast Guard, and the Department of the Interior, came to Ottawa and met with the Secretary of State for External Affairs, the Minister of Land, the President of the Privy Council and senior Canadian officials, including the Under Secretary of State for External Affairs, and Canada's Ambassador to Washington, to make known the USA views on the questions under discussion. These discussions lasted all day and were again frank but friendly. Subsequently there were further discussions in Washington between our Ambassador to the USA and Mr Johnson, and a phone conversation between the Prime Minister and Secretary of State Rogers. Unfortunately, it did not prove possible for the two Governments to reach agreement on all aspects of these questions, as has since been made known by the USA Government. I think this account of these discussions makes quite clear that we have taken very seriously the USA interest in these matters. These differences can be resolved, and resolved in a manner consistent with our interests as a sovereign nation and our long history of close and mutually co-operative relations with the USA. We cannot abdicate our responsibilities in a matter of special importance to us, and we cannot abandon our right and duty to protect our territory. Given this fundamental and irreversible position on our part, there remain nevertheless a wide range of possibilities for bilateral and multilateral co-operation which could advance the cause of environmental preservation in the Arctic waters in harmony with the interests of all concerned. We are prepared to go forward from this position, but only forward and not back.

2nd Session, 28th Parliament, 18–19 Elizabeth II, 1969–70

THE HOUSE OF COMMONS OF CANADA

BILL C-202

An Act to prevent pollution of areas of the arctic waters adjacent to the mainland and islands of the Canadian arctic

Whereas Parliament recognizes that recent developments in relation to *Preamble* the exploitation of the natural resources of arctic areas, including the natural resources of the Canadian arctic, and the transportation of those resources to the markets of the world are of potentially great significance to international trade and commerce and to the economy of Canada in particular;

213

And whereas Parliament at the same time recognizes and is determined to fulfil its obligation to see that the natural resources of the Canadian arctic are developed and exploited and the arctic waters adjacent to the mainland and islands of the Canadian arctic are navigated only in a manner that takes cognizance of Canada's responsibility for the welfare of the Eskimo and other inhabitants of the Canadian arctic and the preservation of the peculiar ecological balance that now exists in the water, ice and land areas of the Canadian arctic;

Now therefore, Her Majesty, by and with the advice and consent of the Senate and House of Commons of Canada, enacts as follows:

SHORT TITLE

Short title   **1.** This Act may be cited as the *Arctic Waters Pollution Prevention Act.*

INTERPRETATION

Definitions   **2.** In this Act,

"Analyst"   (*a*) "analyst" means a person designated as an analyst pursuant to the *Canada Water Act* or the *Northern Inland Waters Act*;

"Icebreaker"   (*b*) "icebreaker" means a ship specially designed and constructed for the purpose of assisting the passage of other ships through ice;

"Owner"   (*c*) "owner" in relation to a ship, includes any person having for the time being, either by law or by contract, the same rights as the owner of the ship as regards the possession and use thereof;

"Pilot"   (*d*) "pilot" means a person licensed as a pilot pursuant to the *Canada Shipping Act*;

"Pollution prevention officer"   (*e*) "pollution prevention officer" means a person designated as a pollution prevention officer pursuant to section 14;

"Ship"   (*f*) "ship" includes any description of vessel or boat used or designed for use in navigation without regard to method or lack of propulsion;

"Shipping safety control zone"   (*g*) "shipping safety control zone" means an area of the arctic waters prescribed as a shipping safety control zone by order of the Governor in Council made under section 11; and

"Waste"   (*h*) "waste" means

(i) any substance that, if added to any waters, would degrade or alter or form part of a process of degradation or alteration of the quality of those waters to an extent that is detrimental to their use by man or by any animal, fish or plant that is useful to man, and

(ii) any water that contains a substance in such a quantity or concentration, or that has been so treated, processed or changed, by heat or other means, from a natural state that it would, if added to any waters, degrade or alter or form part of a process of degradation or alteration of the quality of those waters to an extent that is detrimental to their use by man or by any animal, fish, or plant that is useful to man,

and without limiting the generality of the foregoing, includes anything that, for the purposes of the *Canada Water Act*, is deemed to be waste.

APPLICATION OF ACT

3. (1) Except where otherwise provided, this Act applies to the waters (in this Act referred to as the "arctic waters") adjacent to the mainland and islands of the Canadian arctic within the area enclosed by the sixtieth parallel of north latitude, the one hundred and forty-first meridian of longitude and a line measured seaward from the nearest Canadian land a distance of one hundred nautical miles; except that in the area between the islands of the Canadian arctic and Greenland, where the line of equidistance between the islands of the Canadian arctic and Greenland is less than one hundred nautical miles from the nearest Canadian land, there shall be substituted for the line measured seaward one hundred nautical miles from the nearest Canadian land such line of equidistance. *Application to arctic waters*

(2) For greater certainty, the expression "arctic waters" in this Act includes all waters described in subsection (1) and, as this Act applies to or in respect of any person described in paragraph (*a*) of subsection (1) of section 6, all waters adjacent thereto lying north of the sixtieth parallel of north latitude, the natural resources of whose subjacent submarine areas Her Majesty in right of Canada has the right to dispose of or exploit, whether the waters so described or such adjacent waters are in a frozen or a liquid state, but does not include inland waters. *Idem*

DEPOSIT OF WASTE

4. (1) Except as authorized by regulations made under this section, no person or ship shall deposit or permit the deposit of waste of any type in the arctic waters or in any place on the mainland or islands of the Canadian arctic under any conditions where such waste or any other waste that results from the deposit of such waste may enter the arctic waters. *Prohibition*

(2) Subsection (1) does not apply to the deposit of waste in waters that form part of a water quality management area designated pursuant to the *Canada Water Act* if the waste so deposited is of a type and quantity and is deposited under conditions authorized by regulations made by the Governor in Council under paragraph (*a*) of subsection (2) of section 16 of that Act with respect to that water quality management area. *Application of subsection (1)*

(3) The Governor in Council may make regulations for the purposes of this section prescribing the type and quantity of waste, if any, that may be deposited by any person or ship in the arctic waters or in any place on the mainland or islands of the Canadian arctic under any conditions where *Regulations*

215

such waste or any other waste that results from the deposit of such waste may enter the arctic waters, and prescribing the conditions under which any such waste may be so deposited.

5. (1) Any person who

(*a*) has deposited waste in violation of subsection (1) of section 4, or

(*b*) carries on any undertaking on the mainland or islands of the Canadian arctic or in the arctic waters that, by reason of any accident or other occurrence, is in danger of causing any deposit of waste described in that subsection otherwise than of a type, in a quantity and under conditions prescribed by regulations made under that section,

shall forthwith report the deposit of waste or the accident or other occurrence to a pollution prevention officer at such location and in such manner as may be prescribed by the Governor in Council.

(2) The master of any ship that has deposited waste in violation of subsection (1) of section 4, or that is in distress and for that reason is in danger of causing any deposit of waste described in that subsection otherwise than of a type, in a quantity and under conditions prescribed by regulations made under that section, shall forthwith report the deposit of waste or the condition of distress to a pollution prevention officer at such location and in such manner as may be prescribed by the Governor in Council.

6. (1) The following persons, namely:

(*a*) any person who is engaged in exploring for, developing or exploiting any natural resource on any land adjacent to the arctic waters or in any submarine area subjacent to the arctic waters,

(*b*) any person who carries on any undertaking on the mainland or islands of the Canadian arctic or in the arctic waters, and

(*c*) the owner of any ship that navigates within the arctic waters and the owner or owners of the cargo of any such ship,

are respectively liable and, in the case of the owner of a ship and the owner or owners of the cargo thereof, are jointly and severally liable, up to the amount determined in the manner provided by regulations made under section 9 in respect of the activity or undertaking so engaged in or carried on or in respect of that ship, as the case may be,

(*d*) for all costs and expenses of and incidental to the taking of action described in subsection (2) on the direction of the Governor in Council, and

(*e*) for all actual loss or damage incurred by other persons
resulting from any deposit of waste described in subsection (1) of section 4 that is caused by or is otherwise attributable to that activity or undertaking or that ship, as the case may be.

(2) Where the Governor in Council directs any action to be taken by or on behalf of Her Majesty in right of Canada to repair or remedy any condition that results from a deposit of waste described in subsection (1), or to reduce or mitigate any damage to or destruction of life or property that results or may reasonably be expected to result from such deposit of waste, the costs and expenses of and incidental to the taking of such action, to the extent that such costs and expenses can be established to have been reasonably incurred in the circumstances, are, subject to this section, recoverable by Her Majesty in right of Canada from the person or persons described in paragraph (*a*), (*b*) or (*c*) of that subsection, with costs, in proceedings brought or taken therefor in the name of Her Majesty. *Costs and expenses of Her Majesty*

(3) All claims pursuant to this section against a person or persons described in paragraph (*a*), (*b*) or (*c*) of subsection (1) may be sued for and recovered in any court of competent jurisdiction in Canada, and all such claims shall rank *pari passu* up to the limit of the amount determined in the manner provided by regulations made under section 9 in respect of the activity or undertaking engaged in or carried on by the person or persons against whom the claims are made, or in respect of the ship of which any such person is the owner or of all or part of whose cargo any such person is the owner. *Procedure for recovery of claims*

(4) No proceedings in respect of a claim pursuant to this section shall be commenced after two years from the time when the deposit of waste in respect of which the proceedings are brought or taken occurred or first occurred, as the case may be, or could reasonably be expected to have become known to those affected thereby. *Limitation period*

7. (1) The liability of any person pursuant to section 6 is absolute and does not depend upon proof of fault or negligence, except that no person is liable pursuant to that section for any costs, expenses or actual loss or damage incurred by another person whose conduct caused any deposit of waste described in subsection (1) of that section, or whose conduct contributed to any such deposit of waste, to the degree to which his conduct contributed thereto, and nothing in this Act shall be construed as limiting or restricting any right of recourse or indemnity that a person liable pursuant to section 6 may have against any other person. *Nature and extent of liability*

(2) For the purposes of subsection (1), a reference to any conduct of "another person" includes any wrongful act or omission by that other person or by any person for whose wrongful act or omission that other person is by law responsible. *Idem*

(3) Notwithstanding anything in this Act, no person is liable pursuant to section 6, either alone or jointly and severally with any other person or persons, by reason only of his being the owner of all or any part of the cargo of a ship if he can establish that the cargo or part thereof of which he is the owner is of such a nature, or is of such a nature and is carried in such a quantity that, if it and any other cargo of the same nature that is *Limitation on liability of cargo owner*

217

carried by that ship were deposited by that ship in the arctic waters, the deposit thereof would not constitute a violation of subsection (1) of section (4).

Evidence of financial responsibility to be provided
8. (1) The Governor in Council may require

(*a*) any person who engages in exploring for, developing or exploiting any natural resource on any land adjacent to the arctic waters or in any submarine area subjacent to the arctic waters,

(*b*) any person who carries on any undertaking on the mainland or islands of the Canadian arctic or in the arctic waters that will or is likely to result in the deposit of waste in the arctic waters or in any place under any conditions where such waste or any other waste that results from the deposit of such waste may enter the arctic waters,

(*c*) any person, other than a person described in paragraph (*a*), who proposes to construct, alter or extend any work or works on the mainland or islands of the Canadian arctic or in the arctic waters that, upon completion thereof, will form all or part of an undertaking described in paragraph (*b*), or

(*d*) the owner of any ship that proposes to navigate or that navigates within any shipping safety control zone specified by the Governor in Council and, subject to subsection (3) of section 7, the owner or owners of the cargo of any such ship,

to provide evidence of financial responsibility, in the form of insurance or an indemnity bond satisfactory to the Governor in Council, or in any other form satisfactory to him, in an amount determined in the manner provided by regulations made under section 9.

Persons entitled to claim against insurance or bond
(2) Evidence of financial responsibility in the form of insurance or an indemnity bond shall be in a form that will enable any person entitled pursuant to section 6 to claim against the person or persons giving such evidence of financial responsibility to recover directly from the proceeds of such insurance or bond.

Regulations respecting manner of determining limit of liability
9. The Governor in Council may make regulations for the purposes of section 6 prescribing, in respect of any activity or undertaking engaged in or carried on by any person or persons described in paragraph (*a*), (*b*) or (*c*) of subsection (1) of section 6, or in respect of any ship of which any such person is the owner or of all or part of whose cargo any such person is the owner, the manner of determining the limit of liability of any such person or persons pursuant to that section, which prescribed manner shall, in the case of the owner of any ship and the owner or owners of the cargo thereof, take into account the size of such ship and the nature and quantity of the cargo carried or to be carried by it.

PLANS AND SPECIFICATIONS OF WORKS

10. (1) The Governor in Council may require any person who proposes to construct, alter or extend any work or works on the mainland or islands of the Canadian arctic or in the arctic waters that, upon completion thereof, will form all or part of an undertaking the operation of which will or is likely to result in the deposit of waste of any type in the arctic waters or in any place under any conditions where such waste or any other waste that results from the deposit of such waste may enter the arctic waters, to provide him with a copy of such plans and specifications relating to the work or works as will enable him to determine whether the deposit of waste that will or is likely to occur if the construction, alteration or extension is carried out in accordance therewith would constitute a violation of subsection (1) of section 4. *Plans and specifications to be provided*

(2) If, after reviewing any plans and specifications provided to him under subsection (1) and affording to the person who provided those plans and specifications a reasonable opportunity to be heard, the Governor in Council is of the opinion that the deposit of waste that will or is likely to occur if the construction, alteration or extension is carried out in accordance with such plans and specifications would constitute a violation of subsection (1) of section 4, he may, by order, either *Powers of Governor in Council*

(*a*) require such modifications in those plans and specifications as he considers to be necessary, or

(*b*) prohibit the carrying out of the construction, alteration or extension.

SHIPPING SAFETY CONTROL ZONES

11. (1) Subject to subsection (2), the Governor in Council may, by order, prescribe as a shipping safety control zone any area of the arctic waters specified in the order, and may, as he deems necessary, amend any such area. *Prescription of shipping safety control zones*

(2) A copy of each order that the Governor in Council proposes to make under subsection (1) shall be published in the *Canada Gazette*; and no order may be made by the Governor in Council under subsection (1) based upon any such proposal except after the expiration of sixty days following publication of the proposal in the *Canada Gazette*. *Publication of proposed orders*

12. (1) The Governor in Council may make regulations applicable to ships of any class or classes specified therein, prohibiting any ship of that class or of any of those classes from navigating within any shipping safety control zone specified therein *Regulations relating to navigation in shipping safety control zones*

(*a*) unless the ship complies with standards prescribed by the regulations relating to

(i) hull and fuel tank construction, including the strength of materials

219

used therein, the use of double hulls and the subdivision thereof into watertight compartments,

(ii) the construction of machinery and equipment and the electronic and other navigational aids and equipment and telecommunications equipment to be carried and the manner and frequency of maintenance thereof,

(iii) the nature and construction of propelling power and appliances and fittings for steering and stabilizing,

(iv) the manning of the ship, including the number of navigating and look-out personnel to be carried who are qualified in a manner prescribed by the regulations,

(v) with respect to any type of cargo to be carried, the maximum quantity thereof that may be carried, the method of stowage thereof and the nature or type and quantity of supplies and equipment to be carried for use in repairing or remedying any condition that may result from the deposit of any such cargo in the arctic waters,

(vi) the freeboard to be allowed and the marking of load lines,

(vii) quantities of fuel, water and other supplies to be carried, and

(viii) the maps, charts, tide tables and any other documents or publications relating to navigation in the arctic waters to be carried;

(b) without the aid of a pilot, or of an ice navigator who is qualified in a manner prescribed by the regulations, at any time or during any period or periods of the year, if any, specified in the regulations, or without icebreaker assistance of a kind prescribed by the regulations; and

(c) during any period or periods of the year, if any, specified in the regulations or when ice conditions of a kind specified in the regulations exist in that zone.

Orders exempting certain ships (2) The Governor in Council may by order exempt from the application of any regulations made under subsection (1) any ship or class of ship that is owned or operated by a sovereign power other than Canada where the Governor in Council is satisfied that appropriate measures have been taken by or under the authority of that sovereign power to ensure the compliance of such ship with, or with standards substantially equivalent to, standards prescribed by regulations made under paragraph (a) of subsection (1) that would otherwise be applicable to it within any shipping safety control zone, and that in all other respects all reasonable precautions have been or will be taken to reduce the danger of any deposit of waste resulting from the navigation of such ship within that shipping safety control zone.

Certificates evidencing compliance (3) The Governor in Council may make regulations providing for the issue to the owner or master of any ship that proposes to navigate within any shipping safety control zone specified therein, of a certificate evidencing, in the absence of any evidence to the contrary, the compliance of such ship

220

with standards prescribed by regulations made under paragraph (*a*) of subsection (1) that are or would be applicable to it within that shipping safety control zone, and governing the use that may be made of any such certificate and the effect that may be given thereto for the purposes of any provision of this Act.

13. (1) Where the Governor in Council has reasonable cause to believe that a ship that is within the arctic waters and is in distress, stranded, wrecked, sunk or abandoned, is depositing waste or is likely to deposit waste in the arctic waters, he may cause the ship or any cargo or other material on board the ship to be destroyed, if necessary, or to be removed if possible to such place and sold in such manner as he may direct. *Destruction or removal of ships in distress*

(2) The proceeds from the sale of a ship or any cargo or other material pursuant to subsection (1) shall be applied towards meeting the expenses incurred by the Government of Canada in removing and selling the ship, cargo or other material, and any surplus shall be paid to the owner of that ship, cargo or other material. *Application of proceeds of sale*

POLLUTION PREVENTION OFFICERS

14. (1) The Governor in Council may designate any person as a pollution prevention officer with such of the powers set out in sections 15 and 23 as are specified in the certificate of designation of such person. *Appointment*

(2) A pollution prevention officer shall be furnished with a certificate of his designation specifying the powers set out in sections 15 and 23 that are vested in him, and a pollution prevention officer, on exercising any such power shall, if so required, produce the certificate to any person in authority who is affected thereby and who requires him to do so. *Certificate of designation*

15. (1) A pollution prevention officer may, at any reasonable time, *Powers*

(*a*) enter any area, place or premises (other than a ship, a private dwelling place or any part of any area, place or premises other than a ship that is designed to be used and is being used as a permanent or temporary private dwelling place) occupied by any person described in paragraph (*a*) or (*b*) of subsection (1) of section 8, in which he reasonably believes

(i) there is being or has been carried on any activity that may result in or has resulted in waste, or

(ii) there is any waste

that may be or has been deposited in the arctic waters or on the mainland or islands of the Canadian arctic under any conditions where such waste or any other waste that results from the deposit of such waste may enter the arctic waters in violation of subsection (1) of section 4;

(*b*) examine any waste found therein in bulk or open any container found therein that he has reason to believe contains any waste and take samples thereof; and

(c) require any person in such area, place or premises to produce for inspection or for the purpose of obtaining copies thereof or extracts therefrom, any books or other documents or papers concerning any matter relevant to the administration of this Act or the regulations.

Powers in relation to works

(2) A pollution prevention officer may, at any reasonable time,

(a) enter any area, place or premises (other than a ship, a private dwelling place or any part of any area, place or premises other than a ship that is designed to be used and is being used as a permanent or temporary private dwelling place) in which any construction, alteration or extension of a work or works described in section 10 is being carried on; and

(b) conduct such inspections of the work or works being constructed, altered or extended as he deems necessary in order to determine whether any plans and specifications provided to the Governor in Council, and any modifications required by the Governor in Council, are being complied with.

Powers in relation to ships

(3) A pollution prevention officer may

(a) go on board any ship that is within a shipping safety control zone and conduct such inspections thereof as will enable him to determine whether the ship complies with standards prescribed by any regulations made under section 12 that are applicable to it within that shipping safety control zone;

(b) order any ship that is in or near a shipping safety control zone to proceed outside such zone in such manner as he may direct, to remain outside such zone or to anchor in a place selected by him,

(i) if he suspects, on reasonable grounds, that the ship fails to comply with standards prescribed by any regulations made under section 12 that are or would be applicable to it within that shipping safety control zone,

(ii) if such ship is within the shipping safety control zone or is about to enter the zone in contravention of a regulation made under paragraph (b) or (c) of subsection (1) of section 12, or

(iii) if, by reason of weather, visibility, ice or sea conditions, the condition of the ship or its equipment or the nature or condition of its cargo, he is satisfied that such an order is justified in the interests of safety; and

(c) where he is informed that a substantial quantity of waste has been deposited in the arctic waters or has entered the arctic waters, or where on reasonable grounds, he is satisfied that a grave and imminent danger of a substantial deposit of waste in the arctic waters exists,

222

(i) order all ships within a specified area of the arctic waters to report their positions to him, and

(ii) order any ship to take part in the clean-up of such waste or in any action to control or contain the waste.

16. The owner or person in charge of any area, place or premises entered pursuant to subsection (1) or (2) of section 15, the master of any ship boarded pursuant to paragraph (*a*) of subsection (3) of that section and every person found in the area, place or premises or on board the ship shall give a pollution prevention officer all reasonable assistance in his power to enable the pollution prevention officer to carry out his duties and functions under this Act and shall furnish the pollution prevention officer with such information as he may reasonably require. *Assistance to pollution prevention officer*

17. (1) No person shall obstruct or hinder a pollution prevention officer in the carrying out of his duties or functions under this Act. *Obstruction of pollution prevention officer*

(2) No person shall knowingly make a false or misleading statement, either verbally or in writing, to a pollution prevention officer engaged in carrying out his duties or functions under this Act. *False statements*

OFFENCES

18. (1) Any person who violates subsection (1) of section 4 and any ship that violates that subsection is guilty of an offence and liable on summary conviction to a fine not exceeding, in the case of a person, five thousand dollars, and in the case of a ship, one hundred thousand dollars. *Deposit of waste by persons or ships*

(2) Where an offence is committed by a person under subsection (1) on more than one day or is continued by him for more than one day, it shall be deemed to be a separate offence for each day on which the offence is committed or continued. *Continuing offences*

19. (1) Any person who *Additional offences by persons*

(*a*) fails to make a report to a pollution prevention officer as and when required under subsection (1) of section 5,

(*b*) fails to provide the Governor in Council with evidence of financial responsibility as and when required under subsection (1) of section 8,

(*c*) fails to provide the Governor in Council with any plans and specifications required of him under subsection (1) or section 10, or

(*d*) constructs, alters or extends any work described in subsection (1) of section 10

(i) otherwise than in accordance with any plans and specifications provided to the Governor in Council in accordance with a requirement made under that subsection, or with any such plans and specifications as required to be modified by any order made under subsection (2) of that section, or

(ii) contrary to any order made under subsection (2) of that section prohibiting the carrying out of such construction, alteration or extension,

is guilty of an offence and is liable on summary conviction to a fine not exceeding twenty-five thousand dollars.

Additional offences by ships (2) Any ship

(*a*) that navigates within a shipping safety control zone while not complying with standards prescribed by any regulations made under section 12 that are applicable to it within that shipping safety control zone,

(*b*) that navigates within a shipping safety control zone in contravention of a regulation made under paragraph (*b*) or (*c*) of subsection (1) of section 12,

(*c*) that, having taken on board a pilot in order to comply with a regulation made under paragraph (*b*) of subsection (1) of section 12, fails to comply with any reasonable direction given to it by the pilot in carrying out his duties,

(*d*) that fails to comply with any order of a pollution prevention officer under paragraph (*b*) or (*c*) of subsection (3) of section 15 that is applicable to it,

(*e*) the master of which fails to make a report to a pollution prevention officer as and when required under subsection (2) of section 5, or

(*f*) the master of which or any person on board which violates section 17,

is guilty of an offence and is liable on summary conviction to a fine not exceeding twenty-five thousand dollars.

Obstruction of pollution prevention officer, etc. (3) Any person, other than the master of a ship or any person on board a ship, who violates section 17 is guilty of an offence punishable on summary conviction.

Proof of offence by person **20.** (1) In a prosecution of a person for an offence under subsection (1) of section 18, it is sufficient proof of the offence to establish that it was committed by an employee or agent of the accused whether or not the employee or agent is identified or has been prosecuted for the offence, unless the accused establishes that the offence was committed without his knowledge or consent and that he exercised all due diligence to prevent its commission.

Proof of offence by ship (2) In a prosecution of a ship for an offence under this Act, it is sufficient proof that the ship has committed the offence to establish that the act or neglect that constitutes the offence was committed by the master of or any person on board the ship, other than a pollution prevention officer or a pilot taken on board in compliance with a regulation made under para-

224

graph (*b*) of subsection (1) of section 12, whether or not the person on board the ship has been identified; and for the purposes of any prosecution of a ship for failing to comply with any order or direction of a pollution prevention officer or a pilot, any order given by such pollution prevention officer or any direction given by such pilot to the master or any person on board the ship shall be deemed to have been given to the ship.

21. (1) Subject to this section, a certificate of an analyst stating that he has analysed or examined a sample submitted to him by a pollution prevention officer and stating the result of his analysis or examination is admissible in evidence in any prosecution for a violation of subsection (1) of section 4 and in the absence of evidence to the contrary is proof of the statements contained in the certificate without proof of the signature or the official character of the person appearing to have signed the certificate. *Certificate of analyst*

(2) The party against whom a certificate of an analyst is produced pursuant to subsection (1) may, with leave of the court, require the attendance of the analyst for the purposes of cross-examination. *Attendance of analyst*

(3) No certificate shall be received in evidence pursuant to subsection (1) unless the party intending to produce it has given to the party against whom it is intended to be produced reasonable notice of such intention together with a copy of the certificate. *Notice*

22. (1) Where any person or ship is charged with having committed an offence under this Act, any court in Canada that would have had cognizance of the offence if it had been committed by a person within the limits of its ordinary jurisdiction has jurisdiction to try the offence as if it had been so committed. *Jurisdiction in relation to offences*

(2) Where a ship is charged with having committed an offence under this Act, the summons may be served by leaving the same with the master or any officer of the ship or by posting the summons on some conspicuous part of the ship, and the ship may appear by counsel or agent, but if it does not appear, a summary conviction court may, upon proof of service of the summons, proceed *ex parte* to hold the trial. *Service on ship and appearance at trial*

SEIZURE AND FORFEITURE

23. (1) Whenever a pollution prevention officer suspects on reasonable grounds that *Seizure of ship and cargo*

(*a*) any provision of this Act or the regulations has been contravened by a ship, or

(*b*) the owner of a ship or the owner or owners of all or part of the cargo thereof has or have committed an offence under paragraph (*b*) of subsection (1) of section 19,

he may, with the consent of the Governor in Council, seize the ship and

225

its cargo anywhere in the arctic waters or elsewhere in the territorial sea or internal or inland waters of Canada.

Custody　(2) Subject to subsection (3) and section 24, a ship and cargo seized under subsection (1) shall be retained in the custody of the pollution prevention officer making the seizure or shall be delivered into the custody of such person as the Governor in Council directs.

Perishable goods　(3) Where all or any part of a cargo seized under subsection (1) is perishable, the pollution prevention officer or other person having custody thereof may sell the cargo or the portion thereof that is perishable, as the case may be, and the proceeds of the sale shall be paid to the Receiver General or shall be deposited in a chartered bank to the credit of the Receiver General.

Court may order forfeiture　**24.** (1) Where a ship is convicted of an offence under this Act, or where the owner of a ship or an owner of all or part of the cargo thereof has been convicted of an offence under paragraph (*b*) of subsection (1) of section 19, the convicting court may, if the ship and its cargo were seized under subsection (1) of section 23, in addition to any other penalty imposed, order that the ship and cargo or the ship or its cargo or any part thereof be forfeited, and upon the making of such order the ship and cargo or the ship or its cargo or part thereof is or are forfeited to Her Majesty in right of Canada.

Forfeiture of proceeds of sale　(2) Where any cargo or part thereof that is ordered to be forfeited under subsection (1) has been sold under subsection (3) of section 23, the proceeds of such sale are, upon the making of such order, forfeited to Her Majesty in right of Canada.

Redelivery of ship and cargo on bond　(3) Where a ship and cargo have been seized under subsection (1) of section 23 and proceedings that could result in an order that the ship and cargo be forfeited have been instituted, the court in or before which the proceedings have been instituted may, with the consent of the Governor in Council, order redelivery thereof to the person from whom they were seized upon security by bond, with two sureties, in an amount and form satisfactory to the Governor in Council, being given to Her Majesty in right of Canada.

Seized ship, etc. to be returned unless proceedings instituted　(4) Any ship and cargo seized under subsection (1) of section 23 or the proceeds realized from a sale of any perishable cargo under subsection (3) of that section shall be returned or paid to the person from whom the ship and cargo were seized within thirty days from the seizure thereof unless, prior to the expiration of the thirty days, proceedings are instituted in respect of an offence alleged to have been committed by the ship against this Act or in respect of an offence under paragraph (*b*) of subsection (1) of section 19 alleged to have been committed by the owner of the ship or an owner of all or part of the cargo thereof.

(5) Where proceedings referred to in subsection (4) are instituted and, at the final conclusion of those proceedings, a ship and cargo or ship or cargo or part thereof is or are ordered to be forfeited, they or it may, subject to section 25, be disposed of as the Governor in Council directs. *Disposal of forfeited ship*

(6) Where a ship and cargo have been seized under subsection (1) of section 23 and proceedings referred to in subsection (4) have been instituted, but the ship and cargo or ship or cargo or part thereof or any proceeds realized from the sale of any part of the cargo are not at the final conclusion of the proceedings ordered to be forfeited, they or it shall be returned or the proceeds shall be paid to the person from whom the ship and cargo were seized, unless there has been a conviction and a fine imposed in which case the ship and cargo or proceeds may be detained until the fine is paid, or the ship and cargo may be sold under execution in satisfaction of the fine, or the proceeds realized from a sale of the cargo or any part thereof may be applied in payment of the fine. *Return of seized ship, etc. where no forfeiture ordered*

25. (1) The provisions of section 64A of the *Fisheries Act* apply, with such modifications as the circumstances require, in respect of any ship and cargo forfeited under this Act as though the ship and cargo were, respectively, a vessel and goods forfeited under subsection (5) of section 64 of that Act. *Protection of persons claiming interest*

(2) References to "the Minister" in section 64A of the *Fisheries Act* shall, in applying that section for the purposes of this Act, be read as references to the Governor in Council and the phrase "other than a person convicted of the offence that resulted in the forfeiture or a person in whose possession the vessel, vehicle, article, goods or fish were when seized" shall be deemed to include a reference to the owner of the ship where it is the ship that is convicted of the offence that results in the forfeiture. *Idem*

DELEGATION

26. (1) The Governor in Council may, by order, delegate to any member of the Queen's Privy Council for Canada designated in the order the power and authority to do any act or thing that the Governor in Council is directed or empowered to do under this Act; and upon the making of such an order, the provision or provisions of this Act that direct or empower the Governor in Council and to which the order relates shall be read as if the title of the member of the Queen's Privy Council for Canada designated in the order were substituted therein for the expression "the Governor in Council". *Delegation of powers of the Governor in Council*

(2) This section does not apply to authorize the Governor in Council to delegate any power vested in him under this Act to make regulations, to prescribe shipping safety control zones or to designate pollution prevention officers and their powers, other than pollution prevention officers with only those powers set out in subsection (1) or (2) of section 15. *Limitation*

227

Fines to be paid to Receiver General

27. All fines imposed pursuant to this Act belong to Her Majesty in right of Canada and shall be paid to the Receiver General.

COMING INTO FORCE

Commence-ment

28. This Act shall come into force on a day to be fixed by proclamation.

INTERNATIONAL WILDFOWL RESEARCH BUREAU

Convention on Wetlands of International Importance, especially as Waterfowl Habitat

Covering Note

After long and intensive study, the text for this Convention has reached its final version. The need for such a convention was felt in 1962 at the International Conference on Wetlands (the MAR Conference), organised jointly by the International Union for Conservation of Nature and Natural Resources (IUCN), the International Council for Bird Preservation (ICBP) and the International Wildfowl Research Bureau (IWRB), and held in les Saintes Maries de la Mer, France, in November 1962. There, it was decided that IUCN should compile a list in accordance with an internationally agreed classification of European and North African wetlands of international importance, together with detailed information on these areas. It was recommended at the same time that this list be considered as a basis for an international convention on wetlands.

In the recommendation of the First European Meeting on Wildfowl Conservation at St Andrews, Scotland, in 1963, the need for an international convention was stressed and the IWRB sent out eight draft points to thirty-five countries. On the reactions received and the opinions expressed at the Second European Meeting on Wildfowl Conservation, at Noordwijk, the Netherlands, in 1966, the Dutch Ministry of Culture, Recreation and Social Welfare started to draft a convention on the conservation of wetlands.

This draft was discussed informally during a working session of the IWRB Executive Board in Morges, Switzerland, in 1967. The formal Draft Convention was circulated prior to the International Regional Meeting on Conservation of Wildfowl Resources in Leningrad in September 1968, in the resolutions of which it was thought "expedient to accelerate the adoption of a convention concerning wetlands conservation, and to provide for strict protection of those wetlands which are of international importance". The Soviets expressed their views on a Wetland Convention in an alternative text.

During the Annual Board Meeting of the IWRB in Vienna in May 1969 both texts were compared and it was decided that a compromise text should be drawn up by the Dutch Ministry. This new text was discussed in March 1970 in Espoo, Finland, by representatives of the international organisations mentioned earlier, together with the Food and Agriculture Organisation (FAO) and the Conseil International de la Chasse (CIC), ecological and legal experts of ten different nationalities, and resulted in the Final Text as presented here. The Final Text of this Convention is therefore the fruit of eight years of careful study and international co-operation among scientists and legal experts of a great number of countries, and international organisations like FAO, IUCN, ICBP, CIC and IWRB.

The evolution of the text of this Convention may appear long drawn out. However, experience with other Conventions (such as the Paris Convention on the Protection of Birds, 1950) has demonstrated the necessity of having prior agreement on a realistic text, if it is to be accepted and acted upon by governments.

The Intention of the Convention

In the discussions outlined above, it became abundantly clear that States would not accept a Convention that infringed their sovereign rights to deal with their own natural resources. It was therefore out of the question to draw up a Convention prohibiting absolutely the change in ecological status of wetlands, backed by mandatory sanctions.

Instead, this Convention seeks to lead States to make declarations of intent to safeguard their wetlands and to do so in concert with other States. Again, as a matter of practicality, this Convention seeks to concentrate conservation effort on the most important wetlands within a State. These are those whose loss would be felt by the international community. Especial emphasis is laid on wetlands vital to the migratory waterfowl since the future of these animals is very much an international responsibility. This emphasis should not lead to any impression that the general ecological value of wetlands and their importance to other animals, to plants, to science, culture and recreation can be ignored.

Another feature of this Convention is that its provisions are in general terms and not spelt out in detail. The detailed information is, however, available through the international organisations as mentioned in the following Explanatory Notes.

This Convention should be considered as supplementing and extending other Conventions concerned with the conservation of the Biosphere, such as

1950 International Convention for the Protection of Birds

1962 International Convention for the Prevention of Pollution of the Sea by Oil

1968 African Convention for the Conservation of Nature and Natural Resources

Explanatory Notes

These seek to answer questions that will probably arise during a consideration of the Convention Text and to set out detailed information that could not properly be included in that Text.

Article 1

1/1. The definition covers any inland water, other than those deeper than six metres, but also intertidal zones and the shallow zones of tideless seas.

1/2. The groups of birds which are especially to be considered are the orders

Gaviiformes	(Divers)
Podicipediformes	(Grebes)
Pelicaniformes	(Pelicans, Cormorants, Darters)
Ciconiiformes	(Herons, Bitterns, Storks, Ibises, Flamingoes)
Anseriformes	(Screamers, Swans, Geese, Ducks)
Gruiformes	(Cranes, Rails)
Charadriiformes	(Waders, Gulls, Terns)

Article 2

2/1. It is to be emphasised that the proposal of a wetland for inclusion on the List, and the extent of its boundaries, is solely the responsibility of the Contracting Party in whose territory it lies.

Decisions on the extent of the riparian or coastal zone and on whether the interior of a large island should be considered within the wetland boundaries should be based on the requirements for the adequate conservation of the wetland itself. These may vary from the avoidance of disturbance to the maintenance of the water-table.

2/2. Guidance on the international significance of a wetland is available from the IWRB and the IUCN which set up three specialised ecological Projects, MAR, Aqua and Telma.

Project MAR has published (1965) *A List of European and North African Wetlands of International Importance* and this has been extended and kept up to date on file. Because the necessary information was more readily available, the value of wetlands was largely measured by their usage by waterfowl – which thus served as "indicators" of the wetlands' biological importance. Information on other zoological groups, on the wetlands' botany and on their general ecology is being steadily collected and may indicate additional wetlands to be of international importance. Primary emphasis was laid on shallow waters since these are the most immediately threatened by technological developments.

Project Aqua has published a provisional *List of Aquatic Sites Proposed for Conservation on the Basis of their Scientific Value*. Here, the criteria are limnological and hydrological interest. Some of these waters are much deeper than six metres. Nevertheless, it is thought appropriate for waters such as these to be inscribed in the List of Wetlands of International Importance and to be afforded the protection of the present Convention. Deep, salt waters, that is the open sea, are specifically excluded since they are the subjects of other conventions.

Project Telma is preparing for publication a list of temperate peatlands which could also be appropriately covered by the present Convention.

2/5. Although provision is made for a Contracting Party to delete or restrict wetlands it has inscribed on the List it is emphasised that this should only be done as a matter of urgent national interest and after informing the other Parties to the Convention.

2/6. The migratory stocks of waterfowl are wholly dependent on the existence of adequate wetlands in which to breed, to pause on migration and to spend the winter. The wetlands must not only be adequate in size but be sufficiently close to serve as staging posts on the migratory routes. It is for these reasons that the international responsibilities for the conservation and rational management of migratory stocks of waterfowl are emphasised.

The question of international agreement on the co-ordination and control of the hunting of waterfowl will, it is hoped, be the subject of a further Convention.

Article 3

3/1. Guidance on the nature of policies to ensure the conservation and rational management of wetlands is to be found in the following publications:

Proceedings of the MAR Conference, Saintes Maries de la Mer, France, 12–16 November 1962 (IUCN Publication No. 3), in English and French

Proceedings of the First European Conference on the Conservation of Wildfowl, St Andrews, Scotland, 16–18 November 1963 (Published by the Nature Conservancy in collaboration with the IWRB), in English and French

Proceedings of the Second European Conference on the Conservation of Wildfowl, Noordwijk aan Zee, the Netherlands, 9–14 May 1966 (Published by the Ministry of Cultural Affairs, Recreation and Social Welfare in collaboration with the State Institution for Nature Conservation Research (RIVON) and the IWRB), in English or French

Proceedings of a Technical Meeting on Wetland Conservation, Ankara, Bursa, Istanbul, 9–16 October 1967 (IUCN Publication No. 12), in English

Proceedings of the International Regional Meeting on Conservation of Wildfowl Resources, 25–30 September 1968, Leningrad (to be published in 1970 by the Ministry of Agriculture of the USSR), in English

Liquid Assets (Published in 1965 by IUCN and the IWRB), in English or French

3/2. In view of the rapidity with which modern technology can, deliberately or accidentally, change the character of wetlands, emphasis is laid on the rapid detection of such changes and the informing of interested parties.

Article 4

4/1. The creation of statutory nature reserves is a vital part of conservation.

4/2. This is especially so where internationally important wetlands are destroyed or diminished. Where possible, the reserves should be in the same area so as not to disrupt migratory patterns, but the setting aside of replacement areas elsewhere may be the only feasible course.

4/3, 4, 5. Conservation can only succeed if based on adequate scientific knowledge and enlightened management. These paragraphs indicate the steps needed to ensure such backing.

Article 5

The need for international accord is particularly acute in the circumstances covered by this Article.

Article 6

It is envisaged that one of the main ways in which the spirit and letter of the

Convention will be implemented is by the gathering together of the Contracting Parties in periodical Conferences. These Conferences should continue, and have a similar nature to, the series held in:

St Andrews, Scotland	1963
Noordwijk, the Netherlands	1966
Leningrad, USSR	1968
Ramsar, Iran	1971

It is anticipated that the time interval between Conferences will be of a similar order and that the existence of the Convention will not therefore increase the number of international conferences nor the expenses to be met by the Contracting Parties.

The relevant international organisations are at present:
International Wildfowl Research Bureau
International Union for Conservation of Nature and Natural Resources
International Council for Bird Preservation
International Union of Game Biologists
Conseil International de la Chasse
Food and Agriculture Organisation of the United Nations
United Nations Educational, Scientific, and Cultural Organisation

Article 7

While representatives should preferably be nationals of their Contracting Party this is not mandatory and one person, duly accredited, could represent more than one Contracting Party, especially where wetlands are shared.

Article 8

While it is reasonable to expect that Contracting Parties shall meet the expenses of their own representatives, the question of general conference expenses must be open to negotiation. It is to be hoped that the host Contracting Party will bear them, but in some circumstances a pooling of resources or international assistance may be needed.

Article 9

The existence of the Convention will generate a great deal of secretariat work. It is generally thought undesirable for a new and separate Secretariat with its own budget to be established.

It is thought that the continuing bureau duties could initially be carried out by the Depositing Countries acting in concert. Eventually one of the existing international conservation organisations should undertake these duties and be appropriately financed. The IWRB has been suggested as a suitable body but its status may not be legally adequate. It could, however, well act on behalf of the IUCN or, if an intergovernmental agency were deemed appropriate, of UNESCO or the FAO. It is to be hoped that this matter could be resolved by the Ramsar Conference.

Article 10

10/2. It is generally thought desirable, since a non-political subject such as the Conservation of Wetlands is being dealt with, that no State should be prevented from becoming a Party to the Convention. It would therefore seem unnecessary to qualify the word "State". If a qualification is insisted upon, two alternative wordings have been suggested:

(a) Any State in possession of wetlands within its boundaries and wishing to take part in the measures of conservation of waterfowl may become a Party to this Convention

(b) Any Member of the United Nations or of one of the Specialised Agencies or of the International Atomic Energy Agency or Party to the Statute of the International Court of Justice having wetlands within its territory and wishing to share in the conservation of the stock of wildfowl may become a Party to this Convention

The first alternative maintains the desirable, all-embracing nature of the Convention. The second alternative would exclude several countries having important wetlands and would be unacceptable to a number of other countries on political grounds.

While the Convention was drawn up with the Palaearctic Zone (Europe, Western Asia and Northern Africa) in mind, its provisions are such that States in other parts of the world could quite appropriately become Contracting Parties.

10/3. The following countries have been proposed as being most appropriate to receive the depositions and to act in concert initially to carry out the continuing bureau duties:

(a) The Netherlands – being the originators of the Convention

(b) The USSR   – being deeply involved in the evolution of the Convention

(c) Iran   – being the host at the Conference at which the Convention will be presented

In addition, these countries have many wetlands of international importance within their borders and are themselves widely representative of the Palaearctic region.

Articles 11–14

These are formal and self-explanatory.

International Wildfowl Research Bureau
Slimbridge (Glos.) GL2 7BX
United Kingdom   April 1970

Convention on Wetlands of International Importance Especially as Waterfowl Habitat*

The Contracting Parties,

Recognising the interdependence of Man and his environment,

Considering the fundamental ecological functions of wetlands as regulators of water regimes and as habitats supporting flora and fauna, especially waterfowl,

Being convinced that wetlands constitute a resource of great economic, cultural, scientific and recreational value whose loss would be irreparable,

Desiring to stem the progressive encroachment on and loss of such wetlands now and in the future,

Recognising that waterfowl in their seasonal migrations transcend frontiers and so should be regarded as an international resource,

Being confident that the conservation of wetlands, their flora and their fauna can be ensured by combining clear-sighted national policies with co-ordinated international action

Have agreed as follows,

Article 1

1. Within the meaning of the present Convention wetlands are areas of marsh or of water shallower than six metres. These areas may be natural or artificial, permanent or temporary, with water that is fresh, brackish or salt, static or flowing.
2. Within the meaning of the present Convention waterfowl are birds ecologically dependent on wetlands.

Article 2

1. Each Contracting Party shall propose suitable wetlands within its territory for inscription on a List of Wetlands of International Importance, hereinafter referred to as the List and appended to this Convention. The boundaries of such wetlands shall be precisely described or delimited on an accurate map, and should incorporate riparian and coastal zones and islands lying within the wetlands.
2. Wetlands should be selected for the List on account of their international significance in terms of ecology, botany, zoology, limnology or hydrology as this becomes known. In the first instance wetlands of international importance

* This Final Text derives from the international meeting of experts, organised by the International Wildfowl Research Bureau, in Helsinki, March 17–19, 1970.

to waterfowl at any season should be included. Where appropriate, deep fresh or brackish waters and areas of peatlands may be added to the List.

3. A wetland included in the List shall become the subject of the joint concern of the Contracting Parties, without prejudice to the exclusive sovereign rights of the Contracting Party in whose territory it is situated.

4. Each Contracting Party shall submit its first entries for inclusion in the List when signing, ratifying or acceding to the Convention as specified in Article 10. The provisions of the Convention shall be effective with respect to the first entries for the List as from the date on which the Convention enters into force for the Contracting Party concerned, as specified in Article 11.

5. Any Contracting Party proposing to add to the List further wetlands situated within its territory, to extend the boundaries of those wetlands already included by it in the List, or, because of its urgent national interests, to delete or restrict the boundaries of wetlands already included by it in the List shall, at the earliest possible time, inform those responsible for the continuing bureau duties specified in Article 9.

6. Each Contracting Party shall be guided by a sense of its international responsibilities for the conservation and rational management of migratory stocks of waterfowl, both when submitting its first entry for the List and when making changes.

Article 3

1. The Contracting Parties shall design and implement their conservation policy in such a way as to ensure the conservation and rational management of wetlands in their territory, particularly of those included in the List.

2. Each Contracting Party shall ensure that it is informed at the earliest possible time if the ecological character of any wetland in its territory and included in the List has changed, is changing or is likely to change as the result of technological developments, pollution or other human interference. Information on such changes shall be passed to those responsible for the continuing bureau duties specified in Article 9.

Article 4

1. Each Contracting Party shall promote the conservation of wetlands and waterfowl by organising nature reserves on wetlands whether they are included in the List or not.

2. Where a wetland in the List is diminished in the urgent national interest, the Contracting Party concerned shall compensate as much as possible for the consequent loss of wetland resources and shall have an especial responsibility to create ample nature reserves both for the waterfowl and to retain an adequate part of the original habitat in that area or elsewhere in its territories.

3. The Contracting Parties shall encourage research and the exchange of data and publications regarding wetlands, their flora and their fauna.

4. The Contracting Parties shall endeavour through management to improve waterfowl production on and usage of appropriate wetlands.

5. The Contracting Parties shall promote the training of personnel competent to investigate and manage wetlands.

Article 5

The Contracting Parties shall consult with each other on the implementation of obligations arising from the previous Articles especially in the case of a wetland extending over the territories of more than one Contracting Party or where a water system is shared by Contracting Parties. They shall at the same time endeavour to co-ordinate and support present and future policies and regulations concerning the conservation of wetlands, their fauna and their flora.

Article 6

1. The Contracting Parties shall, as the necessity arises, organise, with the help of those responsible for the continuing bureau duties specified in Article 9 and of the relevant international organisations, Conferences on the Conservation of Wetlands and Waterfowl.
2. The Conferences shall have an advisory character and shall be competent *inter alia*:

   a. to discuss the implementation of the present Convention.
   b. to make known their opinions concerning wetlands proposed for inclusion in the List and concerning proposals for changes in the boundaries of such wetlands as referred to in Paragraph 5 of Article 2.
   c. to make recommendations regarding changes in the ecological character of wetlands included in the List and referred to in Paragraph 2 of Article 3.
   d. to make general or specific recommendations to the Contracting Parties regarding the conservation and rational management of wetlands, their flora and fauna.
   e. to request the relevant international bodies to draw up reports and statistics which are essentially international in character.
3. The Contracting Parties when managing their wetlands shall take into consideration as far as possible the opinions, recommendations and resolutions of the Conferences concerning the conservation and rational management of wetlands, their flora and fauna.

Article 7

1. Each Contracting Party should be represented at a Conference by persons who are experts on wetlands and waterfowl by reason of knowledge and experience gained in either a scientific or an administrative capacity, and who are nationals of the Contracting Party.
2. Each of the Contracting Parties represented shall have one vote at a Conference, resolutions being adopted by a simple majority of the votes cast provided that not less than half the Contracting Parties cast votes.

Article 8

Each Contracting Party shall bear the expenses of its own delegations to the Conferences.

Article 9

1. By agreement amongst themselves, the Depositing Countries specified in Article 10, Paragraph 3, will carry out initially the continuing bureau duties imposed by the present Convention. Eventually one of the existing international organisations concerned with conservation will be charged with these duties.

2. The continuing bureau duties shall be, *inter alia*:

    a. to assist in the convening of the Conferences specified in Article 6, and in the drafting of their agenda.

    b. to maintain the List of Wetlands of International Importance and to be informed by the Contracting Parties of any additions, deletions or restrictions concerning wetlands on that List as referred to in Paragraph 5 of Article 2.

    c. to be informed by the Contracting Parties of any changes in the ecological character of wetlands included in the List as referred to in Paragraph 2 of Article 3.

    d. to forward notification of any alterations to the List or changes in character of wetlands inscribed thereon, together with accompanying documents, to all Contracting Parties and to arrange for these matters to be discussed at the next Conference.

    e. to make known to the Contracting Party concerned the opinions, recommendations and resolutions of the Conferences in respect of such alterations to the List or of changes in the character of wetlands thereon.

Article 10

1. The present Convention shall be open for signature indefinitely.

2. A State may become a Party to this Convention by

    a. signature without reservation as to ratification;

    b. signature subject to ratification followed by ratification;

    c. accession.

3. Ratification or accession shall be effected by the deposit of an instrument of ratification or accession with the Governments of........................
....
at...

Article 11

1. The Convention shall enter into force on 1st January of the year following that in which at least seven States have become Parties to the Convention in accordance with Paragraph 2 of Article 10.

2. For States becoming Parties to the Convention after it has entered into force

the Convention shall come into force on 1st January of the year following that in which they have become Parties.

Article 12

1. The Convention shall be concluded for an indefinite period.
2. Any Contracting Party may, by giving written notice to the Government(s) of.. denounce the Convention at any time after a period of five years from the date on which it entered into force with respect to the Contracting Party concerned. Denunciation shall take effect on 1st January of the year following that in which notice thereof is received.

Article 13

1. a. Any Contracting Party responsible for the international relations of a territory, shall take such measures as may be appropriate in an endeavour to extend the present Convention to the territory and may at any time by notification in writing to the Governments of.......................
   ....
   declare that the present Convention shall extend to such territory.
   b. The present Convention shall, from the date of receipt of the notification or from such other date as may be specified in the notification, extend to the territory named therein.
2. a. Any Contracting Party which has made a declaration under subparagraph a. of Paragraph 1 of this Article may, after the expiry of a period of five years from the date on which the Convention has been so extended to any territory, declare that the Convention shall cease to extend to such territory by notification in writing to the Governments of.............
   ....
   b. The present Convention shall cease to extend to any territory mentioned in such notification one year, or such longer period as may be specified therein after the date of receipt of the notification by the Governments of
   ....

Article 14

1. The Governments of...
   shall inform all States that have signed and acceded to the Convention as soon as possible of:
   a. signatures to the Convention
   b. depositions of instruments of ratification of the Convention
   c. depositions of instruments of accession to the Convention
   d. the date of entry into force of the Convention
   e. notifications of denunciation of the Convention
   f. notifications referred to in Paragraphs 1 or 2 of Article 13.

2. When the Convention has entered into force, the Governments of . . . .
. . . .
shall have it registered with the Secretariat of the United Nations in accordance with Article 102 of the Charter.

IN WITNESS WHEREOF, the undersigned, being duly authorized to that effect, have signed the present Convention.

DONE at . . . . this . . . . day of
. . . . 19 . . . .   in the English, French, German and Russian languages, to be deposited with the Governments of . . . .
. . . .
which undertake to send true copies thereof to all Contracting Parties, and, by agreement amongst themselves, to carry out initially the continuing bureau duties imposed by this Convention.